> Transport, Engineering and Architecture

Hugh Collis

Laurence King Publishing in association with Arup

LAURENCE KING

Published in 2003 by Laurence King Publishing Ltd
71 Great Russell Street
London WC1B 3BP
United Kingdom
Tel: +44 20 7430 8850
Fax: +44 20 7430 8880
e-mail: enquiries@laurenceking.co.uk
www.laurenceking.co.uk

Arup
13 Fitzroy Street
London W1T 4BQ
United Kingdom
Tel: + 44 20 7636 1531
Fax: + 44 20 7580 3924
e-mail: corporate@arup.com
www.arup.com

A catalogue record for this book is available from the British Library

ISBN 1 85669 321 X

Designed by Gavin Ambrose

Printed in China

Picture on page 2 shows Bult station, Hanover Light Rail System,
Germany, by Despang Architekten

Contents

> **Introduction**   6

> **Chapter 1: Airports**   46

Terminal 4, John F. Kennedy International Airport   48
New York, USA, 1994–2001

Terminal 2, Cologne/Bonn Airport   60
Cologne/Bonn, Germany, 1993–2000

New Terminal, Lester B. Pearson International Airport   66
Toronto, Canada, 1997–2004

Hong Kong International Airport, Chek Lap Kok   72
Hong Kong, China, 1991–1998

> **Chapter 2: Railways**   88

Hong Kong Airport Railway   90
Hong Kong, China, 1992–1998

Channel Tunnel Rail Link   116
South-east England, UK, 1989–2007

St Pancras Station   126
London, UK, 2001–2007

King's Cross/St Pancras Underground Station   134
London, UK, 2004

Hanging Railway Stations   142
Wuppertal, Germany, 1993–2003

Light Rail Stations   148
Hanover, Germany, 1996–2000

> **Chapter 3: Bridges**   154

Øresund Crossing   156
Denmark/Sweden, 1993–2000

Millennium Bridge   162
London, UK, 1997–2002

Pero's Bridge   172
Bristol, UK, 1993–1998

Spencer Street Footbridge   176
Melbourne, Australia, 1997–1998

Denver Millennium Bridge   182
Colorado, USA, 1999–2002

Corporation Street Footbridge   188
Manchester, UK, 1998–1999

Hulme Arch Bridge   192
Manchester, UK, 1995–1997

> **Chapter 4: Transport Planning
and Special Projects**   198

Transport for the 2000 Olympics   200
Sydney, Australia, 1996–2000

International Border Crossings and Transport   206
Central and Eastern Europe, ongoing since 1993

Traffic Calming Scheme   212
Brooklyn, New York, USA, 1999–2000

CargoLifter Airship Hangar   218
Brand, Brandenburg, Germany, 1997–2000

Millennium Wheel Canal Link   230
Falkirk, Scotland, UK, 1999–2002

> **Appendix**   234

References and Credits   236

Index   238

> Introduction

'What the engineer sees as structure, the architect sees as a sculpture. Actually, of course, it is both.'

Sir Ove Arup

The last 30 years have seen a revival of interest in the design of buildings both to enhance the experience of the traveller and benefit the business of transport. Collaborations between architects and engineers are now merging the science of new materials and construction methods with the art of architecture, resulting in buildings that are a pleasure to use. This book seeks to document the state of the art in transport design at the turn of the century through the work of Arup in co-operation with various architects of vision and imagination. All fields of transport are explored – buildings, infrastructure and operations. The culture of excellence in design in these fields makes for an exciting period in the development of railways, airports and bridges, and the numerous other buildings and structures that are part of modern systems of transport and communication.

Transport systems present particular challenges to building designers, for the brief is not confined to a single site or locality. Transport infrastructure is part of a network that can span a city, a region, a country or the world. The purpose of such a project is to facilitate movement, not to cater for an activity contained within a single building. With a transport building, therefore, unique dimensions, time and distance are involved, which are not present in a residential or office development. The point of arrival in a city, be it a railway station, airport or bridge, is now part of the defining urban landscape.

Collaborations between Arup and leading architects such as Norman Foster and Renzo Piano – on Stansted and Kansai airports – have placed the firm at the forefront of developments in transport design. However, Günther and Martin Despang's Hanover Light Rail System (see page 148) demonstrates that the transport sector also provides opportunities for imaginative architects in small firms to display their skills and develop their reputations.

Ove Arup, the founder of the practice that bears his name, was one of the first designers of the modern era to recognize the value of the synergy between architects and engineers. Since its formation in 1946 the firm has expanded to work on major commissions with leading designers throughout the world. For most building projects today the architect is the leader, but in the 1990s the fusion of engineering and architecture became the natural process in transport design. The separation of the roles of architect and engineer has diminished as the concept of a design team has replaced the previous practice where one or the other was identified as lead designer.

Previous page: Howrah Bridge, Calcutta, India. Built in the Second World War, it carries 60,000 vehicles and 2 million commuters every day.

Opposite (figure 1): Ove Arup's sketched options for use of shell structures, in an article for *Architectural Design*, 1947.

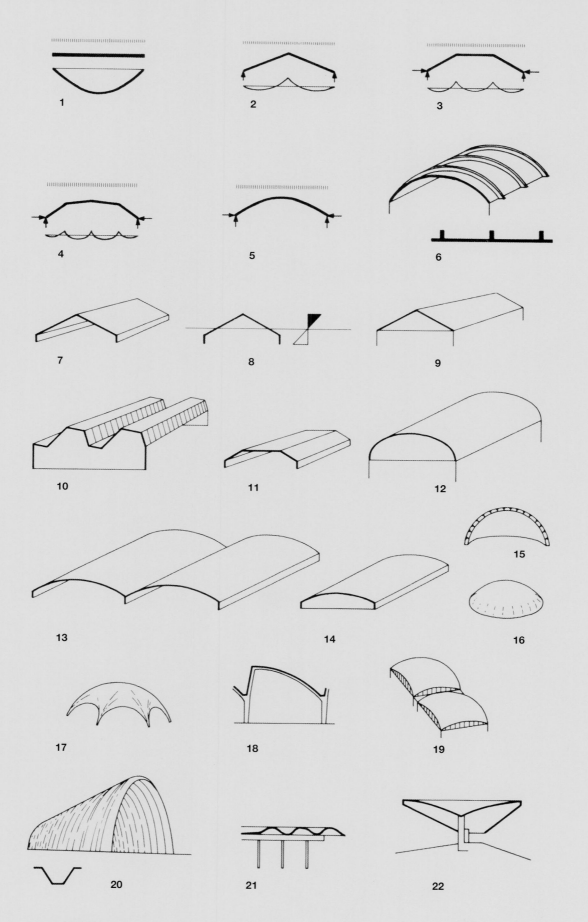

Shell Structures

Transport infrastructure was a natural milieu for the expression of Ove Arup's design philosophy. One of the first commissions for the new firm of Ove Arup & Partners was a bus station for the centre of Dublin, Ireland (architect: Michael Scott). The first significant bridge was added to the firm's portfolio in 1963 with the Kingsgate Footbridge in the historic centre of Durham, UK. When other engineers were looking to drive down the costs of structures by making economies in the use of steel and concrete, Ove Arup was exploring the use of structural materials for reasons of both economy and aesthetics. In an article for *Architectural Design* in 1947, for example, he wrote about concrete shell structures (fig. 1), which as well as being the structural concept for Jørn Utzon's Sydney Opera House (fig. 2) have many applications in transport.

In the 1950s and 1960s the design of most air terminals, railway stations and bridges was purely functional – with the honourable exception of the occasional masterpiece such as Eero Saarinen's TWA Terminal at JFK Airport (then Idlewild), completed in 1962 (fig. 3). John Betjeman, architectural historian and Poet Laureate, in his 1972 book on London's railway terminals, regretted that there was nothing modern worthy of a photograph. Betjeman's comments could have applied to transport architecture in many of the major cities of the world at that time. Post-war austerities and a lack of concern for design quality had resulted in a culture of mediocrity, where engineers were concerned mainly with cost and speed of construction, and architects produced utilitarian and, in many ways, uninspiring, buildings.

Opposite (figure 2): Sydney Harbour Bridge, framing Sydney Opera House, was designed by the British engineer Ralph Freeman and completed in 1932.

Above (figure 3): Eero Saarinen's TWA Terminal at JFK Airport, New York – a rare example of an exciting transport building of the 1960s.

Left (figure 4): The Euston Arch in London, photographed here in 1890, was destroyed in 1961 in a noted act of vandalism by the British Railways Board. The whole station was demolished and replaced by a bleak steel and glass box as part of the West Coast Main Line electrification project.

Right (figure 5): Following the demolition of Penn Station, Jacqueline Kennedy campaigned for a preservation order on New York's Grand Central Station. The restored entrance hall is now one of the most beautiful spaces in Manhattan.

Betjeman's interest in railway architecture was first stimulated by the destruction of Euston Station. Although damaged, the Great Hall, the Shareholders' Meeting Room and the Doric-style Euston Arch (fig. 4) survived the bombing of London during the Second World War only to fall victim to the demolition ball in 1961. Euston wasn't the only fine railway station to be destroyed in a world city: Penn Station in New York and Gare Montparnasse in Paris suffered the same fate in the early 1960s.

The loss of these important buildings and the inadequacy of their replacements created a demand for conservation and quality. In London the campaign was inspired by Betjeman. In New York a very different champion emerged – Jacqueline Kennedy, who had been incensed by the demolition of Penn Station. She went on to play an influential role in the granting of landmark status to Grand Central Station, where imaginative restoration has resulted in Whitney Warren's concourse of 1913 becoming one of the major public spaces in New York today (fig. 5).

Transport, Engineering and Architecture

The concern for the conservation of historic transport buildings has been matched by a desire on the part of both engineers and architects to construct modern buildings of comparable merit. This book seeks to illustrate the fulfilment of that desire.

The late 20th century was not the first period in which transport undertakings engaged the creative skills of architects and designers. In the case of St Pancras in London and Gare d'Orsay in Paris the main role of the architect was the construction of the station hotel. This formed the face of the station in the urban streetscape, and the engineers created the vast train shed behind.

Architects have also played a significant role in creating a unifying identity for transport systems. In Paris, a corporate style was created for the Métro by Hector Guimard in the first decade of the 20th century. This was most visibly expressed in the florid Art Nouveau entrances to the stations (fig.6). On the London Underground, a similar cohesion between stations had to wait until the lines were brought together under the management of Frank Pick in the 1930s. His co-ordinated design framework was reflected in the new stations by Charles Holden and the world-famous London Underground map, Harry Beck's design classic of 1933 (fig. 7).

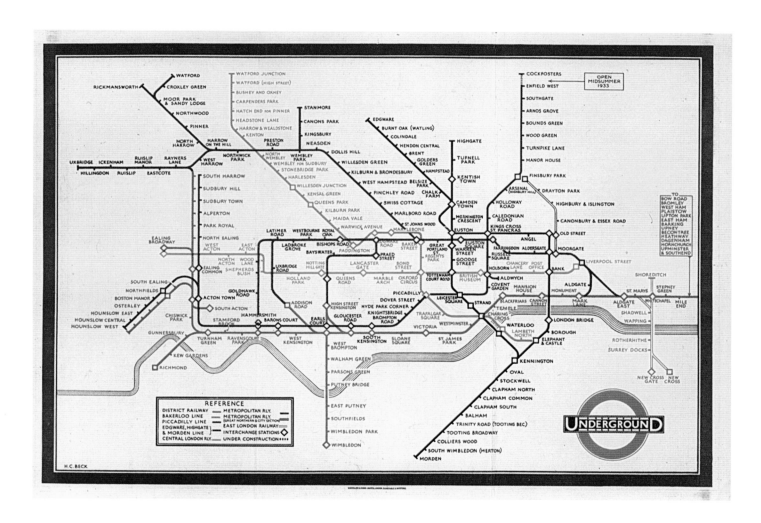

Left (figure 6): Design Classic 1: Hector Guimard designed the distinctive entrances of the Paris Métro in the first decade of the 20th century.

Above (figure 7): Design Classic 2: Harry Beck's London Underground map of 1933 has remained in use ever since, and has proved adaptable to changes in the network.

Above (figure 8): The Pont de Normandie in northern France dominates the landscape at the mouth of the river Seine.

Right (figure 9): The mitre-gate canal lock was invented by Leonardo da Vinci and first used in the 16th century. This lock is near Hungerford, Berkshire.

Opposite (figure 10): Although canal use declined with the growth of the railways, large waterways such as the Panama Canal still have an important role to play in world shipping.

Modern life includes extensive travel, but travelling can be tedious and to many people the time spent travelling is time wasted. Trains and aircraft now have on-board entertainment with music, films and computer games to while away the hours. Enlivening the waiting and interchange experience, and reducing the stress of interchange, is now seen as essential as stations and airports are viewed as part of the total journey experience.

Raising the quality of design at transport facilities also enables operators to realize commercial benefits. Good design is not just an end in itself. Air passengers can often choose which airport to use, and the airport authority benefits from attracting passengers in two ways – through charges to airlines and money spent in its shops and cafés. At busy railway stations waiting times may be shorter, but the higher number of passengers passing through provides an opportunity to attract purchasers of newspapers, snacks and drinks.

The aesthetic dimension has also assumed greater importance in the design of bridges. Larger bridges can express dominance of the environment through their engineering, as demonstrated by the Øresund Crossing (see page 156), the Pont de Normandie in France (fig. 8) and the Tsing Ma Bridge in Hong Kong. The first two dominate the flat landscape which surrounds them, creating visual interest where previously there was none. In Hong Kong, Tsing Ma, one of the largest bridges in the world, appears as a contrasting element in a landscape made up of high buildings and the mountainous islands rising out of the sheltered waters of Victoria Harbour.

Imaginative architects and engineers have employed modern materials and design techniques to transform bridges into sculpture, particularly smaller bridges in high-profile locations. Many bridges have been planned not solely to cross a space, but also to act as an identifying symbol or as functional sculpture to add interest to a locality. For example, London's Millennium Bridge (page 162) forms an essential part of the regeneration of the South Bank area in the centre of the city, and provides a link between two major tourist attractions, the Tate Modern art gallery and St Paul's Cathedral. Another Millennium Bridge, in Denver (see page 182), was also designed to create new links in a regeneration area. In both these cases the visual impact of the bridge is as important as its transport function.

Growth in the demand for travel

Up until the 18th century almost all personal travel and the transport of goods was by river or sea. Unless they were conscripted into the army most people never left the village in which they were born. They walked to work and marched to war. Until the construction of turnpike roads and the development of canals, inland transport was by horse or river vessels. The canalization of rivers and the construction of new canals was made possible by the use of the pound lock (fig. 9) with its two pairs of mitre gates, as invented by Leonardo da Vinci and first used in the 16th century. The architecture of waterways and turnpikes was domestic and modest in scale, comprising accommodation for toll collectors and lock keepers, and was normally in the materials and style of the neighbourhood. It was therefore in the construction of churches and cathedrals that advances in building technology took place.

The Industrial Revolution, which had its origins in Britain, created both the demand for more efficient transport and the means of satisfying it. Iron rails allowed horse-drawn wagons to carry raw materials from mines and quarries to the canals, which as well as providing a system for the distribution of manufactured goods over a wide area, brought coal and iron ore together for smelting. The ability to smelt and distribute large quantities of iron made possible the construction of railways and locomotives. The first commercial steam railway opened between Stockton and Darlington in 1825.

With the growth of railways many canal systems became redundant, but this was not the end of canal construction. The two great ship canals, Suez and Panama (fig. 10), were constructed in the late 19th century, and the Rhine and Danube rivers were finally connected by canal in the last decade of the 20th century.

Above (figure 11): Early European railways did not have the continental ambitions of their North American counterparts. This Thalys train, at Cologne Central Station, is designed to use the electrification systems of four countries' railways. It travels between European cities at 300 kph (186 mph).

Initially railways were planned for the conveyance of goods, and the carriage of passengers was an afterthought. The early railways were therefore constructed to carry bulk products such as coal and building stone from mines and quarries to the sea or nearest waterway, and any passengers were carried in open mineral wagons. The first era of railway architecture did not really begin until the 1840s, when the potential of passenger traffic was realized and there was consequently a requirement for stations with waiting and refreshment rooms and protection from the weather. With increased growth in the trade in manufactured goods, there was also a need for warehouses to store goods at stations.

Although Britain is seen as the birthplace of steam railways, the technology was soon exported to America, where it was enthusiastically adopted. Many American engineers visited Britain between 1825 and 1840, and the engineering of railways in North America was largely based on British experience. The first American railroads were built for connecting the coalfields of Pennsylvania with navigable waters. Over 5,300 kilometres (3,300 miles) were constructed in the USA in the 1830s, and by 1840 the country could boast half of the world's total. In continental Europe the early growth of railways was slower, with most lines being local and unconnected.

The Belgians were the first to realize the potential of a national rail system, and the first international railway opened between Cologne and Antwerp in 1843 (fig. 11).

The 8,800 kilometres (5,500 miles) of railway in 1840 had grown to 350,000 kilometres (220,000 miles) by 1880, with main lines stretching all across Europe and North America. In the last two decades of the 19th century the extent of world railways doubled again, with rapid growth in Russia, Latin America, Africa and Australia. A period of consolidation followed for the main railway networks, and the focus moved to urban transit systems.

Prior to the Second World War high quality inter-urban roads were constructed in Germany (fig. 12). After the war all developed countries built freeway networks of dual-carriageway roads with junctions grade-separated so that vehicles did not have to cross other traffic streams at intersections (fig. 13). This fundamentally changed the way goods were distributed, enabling lorries to travel quickly and reliably from factory or port to warehouse and from warehouse to end user without having to transfer goods between trucks and trains. The railways retained bulk traffic, but most other types of freight moved to the roads, and the organization of business changed as a result.

Above (figure 12): The first high-speed roads in Europe were built in Germany, to assist the movement of Hitler's armies, as pictured here in 1936.

Below (figure 13): Interchanges have since developed as the volume of traffic has grown, as shown by a more recent autobahn in Germany.

Many activities became more centralized, with large distribution warehouses sited at freeway intersections, and as a result freight tonne-kilometres rose significantly faster than Gross Domestic Product. These warehouses did not, however, excite the interest of imaginative designers. Cost was paramount, and simple single-storey steel sheds protected goods from the weather. Ring roads, originally designed to keep through-traffic out of towns and cities, then became the focus of development, because they were accessible from a wide area via uncongested roads. Soon, however, the ring roads also became congested. This did not prevent the inexorable growth in car and truck use, and buses, trams and railways declined in importance.

The latter half of the 20th century saw the beginning of a new era as jet aircraft rapidly superseded ocean liners as the dominant means of intercontinental travel. The first scheduled air service between Europe and North America started in 1946, between London Heathrow and New York La Guardia. Over the following years falling air fares led to an explosion in leisure travel, initially to short-haul destinations in the 1970s and 1980s, but increasingly across the world in the 1990s. There was also rapid growth in corporate air travel as a result of the increased globalization of business. Despite the Gulf War, which caused a reduction in air use between 1990 and 1992, London's airport traffic grew by 75 per cent in the 1990s. Similarly, the impact of the events of 11 September 2001 is expected to be of short duration. One year on, only flights to, from and within the United States were still significantly affected.

High levels of historical growth have been seen at airports across the world, and industry forecasts expect this trend to continue for the foreseeable future. It is therefore not surprising that many of the world's leading architects and engineers are working on new and expanded airport terminals.

The success of low-cost air travel led to congestion in the skies, however, and as a consequence high-speed trains were developed to compete with aircraft on short-haul routes. In addition, commuting by rail into major cities such as London, Paris and New York increased as workers moved away from city centres to improve their quality of life. This revival in passenger railways has led to opportunities for engineers and architects: new lines require new stations and existing stations in the city centres need to be converted to cater for the new patterns of traffic.

Opposite top (figure 14): Pioneering American railways built modest timber stations. Many were replaced by monumental edifices in the early 20th century.

Opposite bottom (figure 15): Car speeds are controlled by landscaping in this example of a traffic-calmed street in Eindhoven, the Netherlands.

Below (figure 16): Fixed-rail urban transport systems are expensive. In Curitiba, Brazil, a lower-cost alternative is provided by triple articulated diesel buses, running in their own lanes and stopping at stations similar to tram stops.

Transport and the growth of cities

The development of suburban railways enabled cities to expand. Some European cities became great industrial centres, with large workforces in engineering and manufacturing. Many of these cities were very small settlements prior to the railway age, existing as small market towns, or with small-scale industries developing around the canal network.

In established cities such as Paris and London, railway stations were initially constructed around the perimeter. As commuter markets developed, congestion increased in the inner core and by the end of the 19th century underground railways were needed to take passengers into and around the heart of the city.

North American cities, apart from East Coast ports such as New York and Boston, developed on a different model. Cities had not existed before the coming of the railway, and so they grew up around the railway stations. The early station buildings were rudimentary affairs, with timber sheds to house the stationmaster and the telegraph office (fig. 14). It is notable that many of the main railway stations in North America were built in the first 20 years of the 20th century, when the growing cities were developing civic pride and the size and quality of the station made a statement about their importance and status.

Cities that developed after the Second World War, and the growing outer suburbs of pre-war cities, were structured on a totally different, car-based model. Growing affluence in the post-war period led to the motor car becoming a universal aspiration rather than a pastime for

Opposite (figure 17): Some Third World countries cannot afford high-quality public transport systems, and in many cities the roads are congested with cars, rickshaws and bicycles, and small privately owned buses known as jitneys, as shown in this busy street in Dhaka, Bangladesh.

Above (figure 18): The Forth Bridge near Edinburgh in Scotland, completed in 1889, is one of the grandest examples of Victorian engineering. Until new techniques were introduced in the 1990s a permanent team of painters worked to protect it from corrosion.

the leisured classes. Cars replaced buses and trains as the means of mass transportation, and public transport patronage and services declined. As car traffic and congestion grew, it became clear that if cities were to function as healthy and attractive places in which to live and work, public transport would have to be developed (fig. 15). Some countries understood the value of public transport better than others. In many European countries public transport has been subsidized and nurtured by enlightened mayors and city administrations, whereas in the UK, North America and Australia, laissez-faire planning has allowed cites to sprawl and much development has been accessible only by private car. This has led to the decline of inner cities.

The transport infrastructure needs of cities are therefore highly dependent on their historical growth pattern. Changes are difficult, because the pressure of existing development imposes constraints on options for improving existing transport systems, and established patterns of use are slow to change. The decline of certain industries and changes in labour-force requirements has created a demand for city regeneration. New uses are being found for the redundant labour-intensive transport facilities of the past such as docks and railway marshalling yards, which usually covered large areas and were often in key inner city locations. Good architecture can provide a focus for the redevelopment of such areas. In Bilbao, for instance, Frank Gehry's Guggenheim Museum has not only revived the port area, but it has put the whole city on the international tourist map. Changes in land use also require changes in transport provision, and new urban railways and bridges are often an essential part of a regeneration strategy.

The cities of the 'Asian Tiger' countries are relatively wealthy and can afford to invest in urban metro systems. They are under intense growth pressures, and urban roads and railways remain essential for the maintenance of their economic development. Cities such as Hong Kong, Bangkok, Singapore and Seoul have been developing mass transit systems over the last 30 years, and others, such as Shanghai and Beijing, will follow suit. As the standard of living in these places rises, so too will expectations for the passenger environment. Examples such as the Airport Express and Tung Chung mass transit lines in Hong Kong point to a future where quality design will be required as well as an efficient transport system (see page 90).

A less expensive solution to traffic congestion, and one more suitable for many cities, is to improve the quality of bus services. In some South American cities that have followed this strategy, buses are now as attractive and reliable as trams (fig. 16).

In the developing world population pressures are enormous, and major cities are magnets for the poor and dispossessed seeking work and food (fig. 17). Civil war and conflict accelerate the move to the cities. Unfortunately the transport solutions available to wealthy countries are not an option here, and transport remains dependent on ill-maintained and elderly buses, which in turn cause high levels of pollution, especially of particulate matter. The 'Asian Brown Cloud', much discussed at the Johannesburg environmental summit of 2002, is attributed partly to slash and burn logging and partly to the emissions from vehicles and domestic fires in cities such as Calcutta.

The role of civil engineering in transport

The construction of canals saw little innovation in civil engineering techniques. Tunnels were blasted through rock, and bridges were of masonry, brick or timber construction. The main tools were the shovel and the horse and cart, and the work was very labour intensive.

The first important development was in the management techniques involved in the mobilization of large labour forces. Previously on major works, craftsmen and labourers had been employed directly by the client, or the architect/engineer on his behalf. The time pressures of the canal business, where shareholders required completion in a matter of months, were totally different from those of earlier complex projects such as cathedrals, where time-scales had been in decades or centuries. The letting of a section of work to a contractor, based on fixed rates or prices, commenced in Britain in the second half of the 18th century and has formed the basis of civil engineering construction in most parts of the world ever since. In this way canals and railways were constructed with impressive speed, sometimes creating hundreds of kilometres of new infrastructure in two or three years.

When railway construction started in the USA, where labour was in short supply, mechanical civil engineering plant was necessary. The West was won by the railways, as the prairies of the Midwest became the bread basket of the world and the produce was transported rapidly and cheaply to the East Coast ports.

Thomas Telford, the founder of the Institution of Civil Engineers, was responsible for one of the first major road projects in Britain between London and Holyhead. He also pioneered the system of separating the engineer from the contractor. The procedures established by Telford

Above (figure 20): The 50-kilometre (30-mile) tunnel between England and France, completed in 1994, was bored by 800-tonne Tunnel Boring Machines moving up to 1,500 tonnes of earth every hour. The back-up train was 235 metres (770 feet) long.

Opposite (figure 19): The trusses for I. K. Brunel's Royal Albert Bridge at Saltash, UK, completed in 1859, were floated out on barges and inched into position. Arup used a similar technique for the Øresund Crossing (pages 156–161).

Above (figure 21): The first tunnel to be built under water since Roman times was the tunnel under the Thames at Rotherhithe in London. At 400 metres (1,300 feet) long, it took 18 years to build. In 1827 the engineer, Marc Brunel, organized a banquet in the partially completed tunnel to prove that it was safe.

had separate roles for the client, the engineer and the contractor and were embraced by both the engineers and the shareholders of the railway companies. They became the basis of the Institution of Civil Engineers' Conditions of Contract and, as the FIDIC (*Fédération Internationale des Ingénieurs-Conseils*) conditions, have been used world wide on major construction projects ever since. British contractors became dominant in the railway industry and sought contracts throughout Europe and, when work became short, in the rest of the world.

The railways required new types of engineering structures. Their need for shallow gradients and straight alignments meant that more bridges and tunnels were necessary, and they had to resist the forces exerted by heavy locomotives. The availability of cast and wrought iron allowed new types of bridges to be built, and tunnelling techniques benefited from the ability of steam engines to provide compressed air. Bridge construction techniques developed rapidly throughout the 19th century, making possible ever larger spans and massive steel structures such as the Forth Railway Bridge (fig. 18). The great engineers of the Victorian era, men such as Isambard Kingdom Brunel in Britain, Gustave Eiffel in France and John Roebling in North America, readily exploited the opportunities that advances in materials science produced. During the 19th century wrought iron was used extensively (fig. 19), but by the time the construction industry moved to road building in the 20th century, steel and reinforced concrete were the commonest materials.

Information technology and advances in mechanical and electrical engineering revolutionized construction methods. The modern laser-guided Tunnel Boring Machine (TBM) (fig. 20) is light years away from the shield used by Marc Brunel's labourers to dig the world's first underwater tunnel at Rotherhithe in London in the 1820s (fig. 21).

The computer revolutionized engineering design by allowing complex analysis calculations to be made in place of simplified manual calculations. This allowed more complex structures to be analyzed and gave architects unprecedented freedom to realize their dreams in steel, glass and concrete. New materials, such as carbon fibre, laminated timber and plastics are now used structurally as well as decoratively, creating opportunities for art and science to combine in 'total architecture' – the cornerstone of Ove Arup's design philosophy.

Railways and their buildings

The railway builders constructed large train sheds at major stations in order to protect passengers from the weather. Their scale was dictated by the use of steam traction: they had to be high enough to allow the smoke and steam to rise above waiting passengers. The painting of Gare St Lazare in Paris by Claude Monet in 1877 (fig. 22) illustrates the nature of the 19th-century railway station. Station buildings included accommodation for railway staff and, at most stations, a passenger waiting room with a coal fire. Major stations frequently had magnificent hotels and, at those where passengers changed trains, enormous dining rooms. Trains were slow and dirty, and waiting was part of the trip. The wealthier customers expected to wait in style, and to eat in style when they broke their journey. As trains became faster and more frequent, and carried more buffet and restaurant cars, many of these elegant station buildings fell into disuse.

The great stations of London, Euston (1837), Paddington (1838), Waterloo (1848), King's Cross (1852) and St Pancras (1868), were built to accommodate long-distance passengers and freight. Paris, like London, built stations around its periphery. Gare St Lazare, Gare d'Austerlitz, the original Gare Montparnasse, Gare de l'Est and the nearby Gare du Nord were all built in the 1840s.

Other European stations of note include Cologne Hauptbahnhof (built in 1859 but rebuilt after the Second World War following extensive damage), Amsterdam Centraal (1889) and the vast Leipzig Hauptbahnhof (1907–15), which claimed to be the largest station in Europe. The entrance hall at Leipzig, which stretches the full width of its 26 tracks, has recently been developed as a shopping centre (fig. 23). This station was the result of the consolidation in one building of services that had previously used six stations owned by different railway companies.

Opposite (figure 22): Victorian train sheds had to protect passengers from the weather, but also allow smoke and steam to rise out of the way. Monet's painting of Gare St Lazare shows the atmosphere in a way no modern photograph could.

Below (figure 23): Leipzig station was the largest in Europe when completed in 1915. Today the enormous concourse has been turned into a three-storey shopping centre, while trains still run from its 26 platforms.

When railway companies merge and services are consolidated, inevitably some stations become redundant. Sometimes they are put to new use. Two examples of changed use are Central Station in Manchester, now the G-Mex exhibition centre (fig. 24), and Gare d'Orsay in Paris (fig. 25), now the Musée d'Orsay.

In North America, the pattern of development of stations differed from Europe, as cities developed around stations rather than vice versa. The original stations in major cities were often modest structures and were replaced by grand buildings, as cities grew and wished to express civic pride through their stations. Washington DC built a station befitting its status as US capital. Union Station (fig. 26), a white granite building 234 metres (760 feet) long and 105 metres (343 feet) wide, was a new claimant to the title of largest station in the world. Much of this station has also become a shopping mall, but part of it is still in use by the railway.

One of the first and most magnificent was Union Station (fig. 27) in St Louis, Missouri, built between 1893 and 1894. At the time it was said to be the largest station in the world, with 30 tracks and covering 4.5 hectares (11 acres). It was complemented by a hotel which opened in 1895. Train use ceased there in 1978 and the shed is now a shopping mall, although the Grand Hall and Hotel were restored and survive.

Perhaps the most magnificent of all stations in the United States is Grand Central in New York, completed in 1913 on the site of an earlier station. After its recent refurbishment New Yorkers have learnt to appreciate once again the glamour of the railroad station. Manhattan's other main terminal, Penn Station (fig. 28), was demolished in 1964 and the station located underground to release the site for property development. The adjacent US Post Office building, built by the same architects in the same style, is shortly to be converted into a new Penn Station concourse.

Above and below (figure 24): Mergers of railway companies made many stations redundant. Central Station in Manchester, England, is now the G-Mex exhibition centre.

Opposite (figure 25): When the Gare d'Orsay in Paris was no longer required as a railway station, it became an art gallery. Trains still run underneath the building.

Above (figure 26): Once American cities became established the early stations were replaced with monuments expressing civic pride. The largest is probably Union Station in Washington DC.

Below (figure 28): Penn station in New York was demolished in 1964. It is now proposed that the adjacent Farley building, designed in the same style by McKim, Mead & White for the US Postal Sevice, will be adapted into a station.

Opposite (figure 27): The station in St Louis, Missouri, was completed in 1894. After railway uses ceased there in 1978 the building was preserved as a shopping mall.

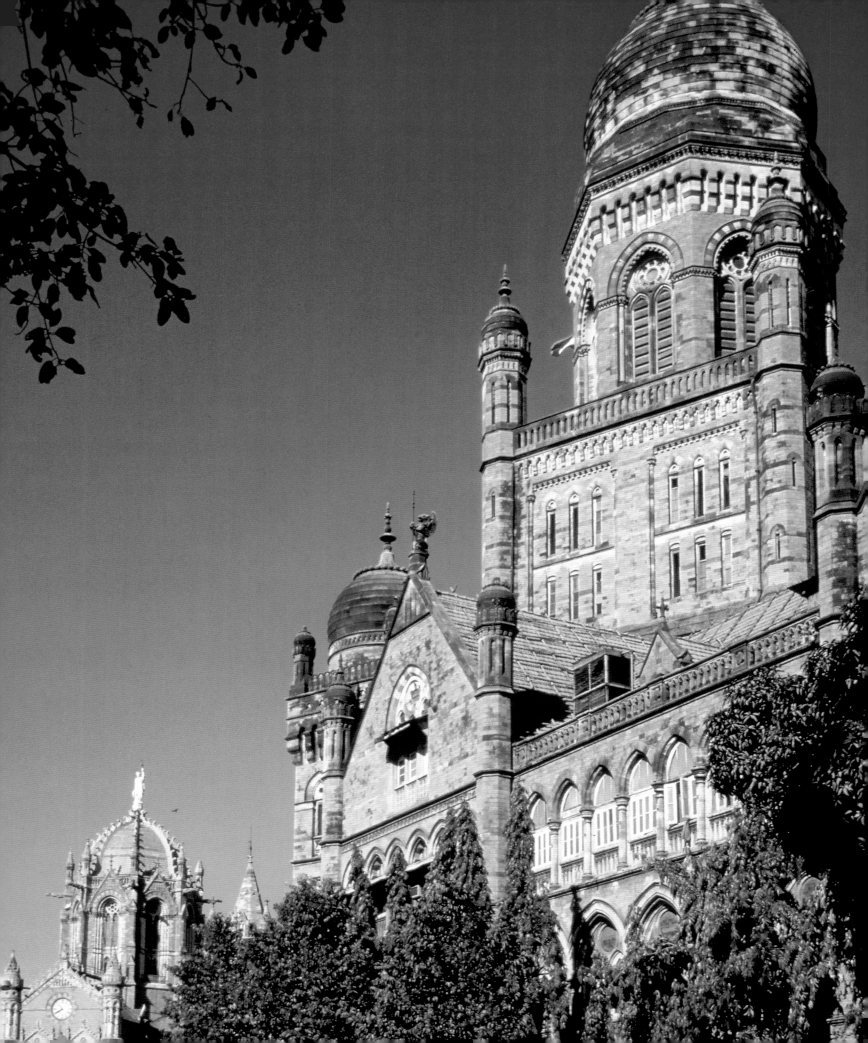

In Asia the main railway stations were a statement made by the ruling colonial power. Two of the most arrogant and ostentatious were Bombay Victoria (recently renamed Mumbai Chhatrapati Sivaji), completed in 1888 (fig. 29), and Kuala Lumpur, completed in 1910 (fig. 30). The station in Bombay was designed by F. W. Stevens, who was appointed in 1877 at the age of 29. He trained as an engineer and joined the Office of the Architect of the Government of India in 1868. Stevens imported the fanciful filigree and carvings of the Italian Gothic style, and later became known as the high priest of Indian Gothic.

The style of the Kuala Lumpur station was also imported. A. B. Hubbock's designs for the booking halls and hotel were Moorish in origin and included towers and minarets as decoration. In 2001 this building was replaced by a new central station, which is contained in the podium for a massive commercial development. The old station is subject to a preservation order and its opulence and charm ensure that it will continue to be one of Kuala Lumpur's main tourist attractions.

In the latter decades of the 20th century, several countries developed high-speed railways for trains operating at speeds of up to 350 kph (220 mph). These new lines normally carry passenger trains only, not heavy freight trains. Such railways are very difficult to construct in areas with high population densities, where choice of alignment is heavily constrained. For example, the Channel Tunnel Rail Link (CTRL) in south-east England (see pages 116–125) is requiring extensive costly tunnelling, and as a result the cost per kilometre is about three times that for the TGV (*train à grande vitesse*) Méditerranée in the more rural south of France (fig. 31). The pioneering high-speed lines in France and Japan have been followed by similar new railways in Germany, Spain, Belgium, the UK and Korea, and studies have been undertaken in the USA and Australia.

Trains now take only three hours to cover the 800 kilometres (500 miles) between Paris Gare de Lyon (rebuilt 1895–1902) and Marseilles Gare St Charles (1848), at speeds of up to 350 kph (220 mph). Historic stations are having to be adapted to accommodate these new types of train. The case study of St Pancras (see pages 126–133) shows the scale of the changes necessary.

Opposite (figure 29): Bombay Victoria station was built by the colonial administration in India, in the Indian Gothic style.

Below (figure 30): Kuala Lumpur station, Malaysia, was completed in 1910 to a Moorish design.

New stations for these new railways – such as Plateau d'Arbois on the TGV Méditerranée and Ebbsfleet on the CTRL – are often placed outside city centres, acting as 'parkway' stations with large car parks. Gare TGV Aix-en-Provence is set in woodland on a wartime military base, serving the suburbs and port areas to the north and west of Marseilles as well as the city of Aix-en-Provence. Ebbsfleet, located near the M25, London's orbital motorway, will have 9,000 parking spaces to attract commuters to London as well as international passengers. Spanish engineer/architect Santiago Calatrava's station at Lyon Satolas airport is set in open countryside on the high-speed line bypassing the city. With its drama and style it has set a new standard that many other stations now seek to emulate.

The need for faster passenger trains is primarily to allow railways to compete with airlines for inter-city traffic on journeys of up to 800 kilometres (500 miles). The travellers attracted to these new trains expect the same quality of passenger comfort and environment in a railway station that they are accustomed to at an airport. Most main railway stations were built for steam trains and are draughty and exposed, so their refurbishment poses a challenge for architects and engineers.

The brief for a new or modernized main railway station usually specifies an airport-style environment, but in practice there are very different functional requirements. Airlines seek to have all their passengers assembled well before departure time in order to minimize loading times and delays. An aircraft carries a fixed number of passengers and air services operate relatively infrequently, so there are problems for passengers who miss flights. Most passengers therefore arrive in good time and have to wait a considerable time in the airport terminal, where extensive shopping, catering and waiting areas are provided to serve them while they kill time between check-in and departure.

Inter-city railways allow passengers to arrive up until just before the train departs, and normally carry passengers on a 'turn up and go' basis as well as those who have pre-booked. Popular routes between major stations usually have at least one train per hour, so a passenger missing a train will not usually have long to wait for the next one and there is no need for extensive lounges. Most major stations also operate local services, which involves handling a large number of passengers in a short period of time. When designing such stations, architects therefore have to plan for clear passageways and direct routes, whereas in airport design they are often required to direct travellers past as many shops and cafés as possible.

Interestingly, rail authorities are now copying airports, and retailing is becoming part of railway-station business. A further development is in-town check-ins for airport express railways. In this way, railway stations and airports are merging into integrated transport terminals, as can be seen in the case of the Hong Kong Airport Express Railway (see pages 90–115).

Opposite (figure 31): The French high-speed train network is the largest in the world. Here, on the TGV Méditerranée, completed in 2001, a 350 kph (220 mph) train crosses the Rhone near Avignon.

Left (figure 32): Richard MacCormac's Southwark station on London's Jubilee Line Extension was completed in 1999. The light, airy booking hall contrasts with the cramped featureless stations of the Victoria Line, completed in 1968.

Increasing congestion in peri-urban and rural areas is leading to a revival of inner-city living, and most urban development policies support this. The Strategic Plan for London, for example, is planning for a population growth of more than 10 per cent over the next 10 years, but foresees a much larger proportionate growth in city-centre employment.

Entirely new urban railway systems have been built or are under construction in cities such as Hong Kong, Singapore and Bangkok. But it is not only the fast-growing cities of the orient that are building new metros. European and North American cities are expanding their existing systems and developing new ones. For example, in Valencia, Spain, a new tunnel has been built across the centre to link two century-old narrow-gauge lines, creating the first metro line. It opened in 1988 and the system is being continually expanded. In Paris a series of cross-city underground railways has been constructed, linking the termini on the fringes of the city, and in London there are advanced plans to do likewise. In New York, which has not had a new subway line for 50 years, work has started on the Second Avenue Subway to increase capacity on the system in Manhattan. Berlin too has been making up for 50 years of neglect, with new and restored metro lines across the city following reunification.

The development of the London Underground after the Second World War featured simple, clean but uninspiring stations on the Victoria Line in the 1960s and on the Jubilee Line in the 1970s. In the 1990s this all began to change, after Roland Paoletti, frustrated by the accountancy and engineering culture he had encountered as architect to the Hong Kong Mass Transit System, was given the opportunity to bring design to the fore with the Jubilee Line Extension, approved in 1991. He appointed a number of leading British architects to design the stations (fig. 32), including Norman Foster, Richard MacCormac, Michael Hopkins, Will Alsop and Ian Ritchie. Even the maintenance depot was not forgotten, for which Chris Wilkinson won many awards.

Stations for urban metro systems are frequently underground, and their layout is thus constrained by both functional requirements and underground obstructions such as sewers and utilities, roads and other railways, and the basements and foundations of buildings. An additional problem is that above-ground sites for access to the station concourses, both for construction and subsequent use, are difficult to find. The case studies on Hong Kong Station (see pages 90–115) and King's Cross/St Pancras in London (see pages 134–141) illustrate the many problems involved in improving and extending metro stations.

Airports

When London's Heathrow Airport opened in 1946 the terminal buildings comprised just a few wartime tents (fig. 33), but by the 1980s new airports and terminals, like railway stations a century before, were blessed with some of the finest architecture of the late 20th century, including prominent works by the world's leading exponents.

Airports provide a tremendous challenge for these designers. They are gateways to cities and seen from the air as well as the surface, and so external appearance is very important. The functional requirements are also highly demanding. Furthermore, the commercial management of airports now requires shopping centres in terminals. These keep waiting passengers occupied and provide revenue for the operator that is subject to less regulation than charges for airline services. In many airports it is now impossible to get from check-in to gate without walking through a shopping mall. BAA, the world's largest airport operator, generates 33 per cent of its revenue through shop rents and concessions, and for Frankfurt Airport the proportion is 25 per cent.

The layout of a terminal requires easy transfer, which is a marketing asset as airports compete for interlining passengers. Attracting travellers to a hub improves the economics of the airlines serving it and creates extra income for the airport authority. Amsterdam Schiphol Airport has long promoted itself as an easy hub to use, and has services from most UK airports, including many that do not have routes to London. Dubai Airport promotes 'Dubai Duty Free', and the new terminal expansion is part of a strategy to develop Dubai as the Middle East's travel and business capital.

The range of surface transport to a major airport can also assist in developing commercial revenues. Heathrow Airport, London, is not only a major European airport for transfer between long- and short-haul flights, but also the biggest coach interchange in Britain, with regular services to most major cities.

Most major airports are growing fast, and expect to continue this growth for the foreseeable future. As a result, most airports have major building works under way most of the time. Scheduling works to maintain adequate operational capacity of terminals and aprons is therefore a daily feature of airport management.

Airports have in some ways recreated the Victorian railway experience with lots of waiting around and breaks in long journeys. Both places are designed for waiting and eating but the two environments couldn't be more different. The great railway stations were built with an expectation of permanence, but airports are much more dynamic. Air services do not rely on fixed tracks, and the demands on an airport can change rapidly as airlines merge or go bankrupt and new air services are developed to cover different markets. As a result, the aviation industry requires buildings with maximum flexibility, a factor which has to be taken into account by the engineers and architects. Like the first railway stations, airport terminals require large, clear spaces – not to accommodate the smoke of trains but to allow for frequent internal rebuilding.

The difficulty in designing for flexibility however is that the uses within the building require a high degree of mechanical and electrical servicing, in addition to normal services such as power, heating and ventilation. Baggage handling requires substantial investment in belts serving check-in desks, baggage sorting and baggage reclaim, and these facilities cannot easily be relocated. Security requirements, which have always been strict, have now assumed even greater importance. The separation of arriving, departing and transferring passengers requires careful planning, for these movements inevitably cross each other in the terminal. Retail and catering spaces need water, sewerage and power services and access for deliveries, yet have to be planned in locations that maximize the commercial benefit to be gained from captive passengers within the airport.

Opposite (figure 33): London's Heathrow Airport, one of the largest international airports in the world, started business in 1946 with army surplus tents for a terminal building.

Below (figure 34): London's Stansted Airport terminal, designed by Norman Foster and completed in 1992, has large clear spaces to provide for flexible use and a modular structure to allow for extension.

Gate lounges, air bridges and other facilities usually have to serve a wide variety of aircraft types, and major airports are now incorporating in their planning the requirements of very large aircraft such as the Airbus A380. Flexibility therefore has to be designed into both the terminal and the apron layouts to allow very different aircraft to be accommodated at the same air bridge at different times during the same day.

The first terminal at Paris Charles de Gaulle (architect: Aeroports de Paris), with its drum shape and long travellators through the building, was considered state of the art when it opened in 1974. However, it proved insufficiently flexible, and Terminal 2 (by the same architects) was designed with the characteristics of linearity and modularity that have become the basis of most terminal designs. Since the first two phases opened in 1981/2 it has been continually expanded, and the sixth module is due for completion in 2003.

Modern airport planning often uses the concept of a 'processor' building, containing check-in, security screening, baggage handling and immigration, and a series of satellites connected by a transit system. Examples are London Stansted (fig. 34), Denver International and the new airport at Bangkok Suvarnabhumi, currently under construction. The satellites, containing the gate lounges and air bridges, allow for aircraft stands on all sides, thus maximizing the flexibility for aircraft manoeuvring on the apron.

As a result of the various constraints imposed on airport design, the most interesting architectural feature often turns out to be the roof (fig. 35). There are two main reasons for this: the roof gives definition and character to the building for passengers arriving both on the surface and by air, and it will normally remain unchanged when internal spaces are redesigned. The soaring roof of Helmut Jahn's Munich Airport Centre Forum (fig. 36) is its defining feature, and the terminal roof of Bilbao Airport was designed by Calatrava to represent a bird in flight.

Opposite (figure 35): The mile-long elegantly arched passenger terminal at Kansai Airport, Osaka, Japan, was designed by Renzo Piano.

Below (figure 36): The soaring Forum roof at the new airport in Munich, Germany, was designed by Helmut Jahn.

Bridges and tunnels

All over the world there are now many crossings between landmasses that traditionally would have been made by boat. The Romans developed the art of constructing masonry bridges, using dressed stones fitted together to create arches. For centuries there was little innovation in bridge construction, and timber and stone remained the main materials. The first iron bridge was constructed across the River Severn at Coalbrookdale, Shropshire, UK, in 1779 (fig. 37). The elements of the bridge were cast by Abraham Darby III at his foundry nearby, and the completed structure had a span of over 30 metres (100 feet) and stood 13.7 metres (45 feet) above the river. The 384 tonne (378 ton) structure, with its five huge cast-iron ribs, still stands today.

The railway age created a huge demand for new bridges in the 19th century. Cast iron was gradually replaced by wrought iron, a less brittle material and thus less subject to fracture as a result of the fatigue stresses imposed by trains. The weight saving also resulted in substantial savings in cost. In 1783 Henry Colt took out the first patents for 'malleable' or wrought iron. Wrought-iron plate girders were first constructed in 1832, the rolled wrought-iron H-beam was patented in 1844, and the tubular beam in 1846. Thomas Telford's Menai Straits suspension bridge in north Wales, completed in 1826, was built with chains made of iron plates. The voracious demands of the railway construction industry meant that all these new technologies were rapidly accepted into common use.

Opposite (figure 37): The Ironbridge at Coalbrookdale, in Shropshire, England, was the first bridge in the world to be made of cast iron.

Above (figure 38): John Roebling's technique for spinning main cables from strands was used for Brooklyn Bridge in New York – and has been a feature of major suspension bridges ever since.

Further innovations followed. In 1855 Henry Bessemer patented his 'Acid Converter' process for making steel which could be rolled into beams and columns, shapes that previously required the riveting together of several iron plates. The American engineer John Roebling was the first to exploit the properties of thin strands by spinning groups of wires into wire ropes to bridge across gaps. His first major bridge, across Niagara Gorge, carried both rail and road ways and was completed in 1855. His greatest work was the Brooklyn Bridge (fig. 38), which was completed in 1883, 14 years after his death, by his son, Washington. It has suspension cables supporting a deck spanning 486 metres (1,595 feet), the longest bridge span in the world at the time. The Clifton Suspension Bridge in Bristol, the greatest bridge work of Britain's most prominent engineer, Isambard Kingdom Brunel, was also completed long after the death of the designer, in 1864. This bridge, spanning 214 metres (702 feet) had chains made from flat plates of wrought iron, originally used for Brunel's Hungerford Bridge across the Thames in London, completed in 1845 but subsequently demolished. Brunel also developed a suspension/arch/truss combination for his railway bridge at Saltash, near Plymouth, which has two spans of 139 metres (455 feet) and was opened in 1859 by Prince Albert.

The major developments in bridge design in the post-war period were in engineering, not in the aesthetic elements. The Severn Bridge in the UK (fig. 39), completed in 1966, was a significant engineering advance, with the use of an aerofoil-shaped steel box girder rather than a truss girder. However, to the untrained eye it follows the same form as all other suspension bridges and its impact stems largely from its position in the landscape.

Today, reinforced and pre-stressed concrete, a wide variety of steels, and new materials such as fibreglass, carbon fibre and plastics, enable the bridge engineer to design structures which can span wide estuaries and sea channels.

The demand for faster travel has led to enormous works of engineering to create fixed crossings to replace ferry services, which as well as being slow are easily disrupted by inclement weather. On the roads and railways, the higher-speed routes have flatter, less curving alignments, and so bridges now have to span valleys that previously would have been crossed with sharp curves and steep gradients. The demands of increasing trade have led to major bridge and tunnel crossings many miles long, like the Storebaelt and Øresund Crossings (see pages 156–162) in Scandinavia, and the links to Japanese islands for both roads and railways (fig. 40).

In recent years it has been recognized that quite modest bridges can contribute interest and focus to both the urban and rural landscape. Architects, artists and sculptors have begun to work with engineers to realize the opportunities created by new materials and the ability to analyse the stresses in quite complex structures. The work of Santiago Calatrava, who often draws on the architecture and engineering of the animal skeleton in his designs, has been a particular inspiration to many designers (fig. 41). Clients have also grasped these opportunities, and bridges in key locations are now frequently the subject of design competitions. Several of the bridges featured in this

book are more important as an icon of urban regeneration than as a transport link, although an identified transport requirement would normally have led to its inception.

Sometimes quite mundane features, such as the counterweight for lifting the main span of Pero's Bridge (see pages 172–175) in Bristol, can be transformed into sculptures that are both artistic and functional. Following the Corbusian edict that 'form follows function' does not require the engineer or architect to produce a traditional, conformist structure. The function of a bridge can often be more broadly defined than merely crossing a gap, and solutions can be exciting and imaginative. In the case of Pero's Bridge, a bascule-type footbridge, a counterweight minimizes the energy required to lift the opening span, but the counterweight is expressed as sculpture. The steel horns created by the sculptor provide the necessary mass to counterbalance the deck. At the same time they create an interesting shape and an attractive feature in an area of historic dockland in the centre of the city. Structures with a wider functionality can create a sense of place or act as a landmark, giving definition to a new development area and generating civic pride as well as raising the spirits. These regenerative functions may be less susceptible to cost-benefit analysis than the saving of journey time resulting from a new bridge, but they are just as real.

The form of the Corporation Street Footbridge in Manchester (see pages 188–192) is designed to obscure the inclination of the deck,

which links two shopping developments with different floor levels. This aesthetic aspect of its functionality is combined with the practical functionality of the glazing, which keeps users of the bridge dry in inclement weather. A central windbreak and canopied roof provide a similar function on the Spencer Street Footbridge in Melbourne (see pages 176–181), adding drama and visual interest. The arch of the Hulme Bridge (see pages 192–197) and the pylon of the Denver Millennium Bridge (see pages 182–187), serve the function of supporting the cables carrying the load of the deck. They also give a sense of place to the redeveloped or regenerated areas in which they are situated.

Marc Mimram's Solférino Bridge across the Seine in Paris took the concept of a level deck supported by an arch a stage further by utilizing the arch for walkways. The pedestrian routes therefore link to both the upper and lower level walkways along the river banks, meeting on the same level in the middle of the river. Like the London

Millennium Bridge (see pages 162–171), Mimram's structure highlighted the risks inherent in designing light, long-span bridges. It had to be closed the day after it was opened in 1999 due to resonant vibrations, and was reopened in 2000 after tuned mass dampers had been fitted.

When he was chosen to design the fourth bridge ever to be built across the Grand Canal in Venice, Calatrava demonstrated that imaginative designers can produce suitable solutions for the most sensitive locations. His design, using timber, is modest and restrained, not seeking to dominate the ambience of one of the world's most famous historic areas.

Bridges have thus become a meeting point for artists, sculptors, engineers and architects who wish to explore the contribution that these new links can make to the urban landscape.

Opposite (figure 39): The Severn Bridge, completed in 1966, connects England and Wales. It was the first major suspension bridge to use an aerofoil shaped box-girder deck.

Above (figure 40): There are always new challenges for the bridge engineer. The Seto Chuo Expressway Route, between Honshu and Shikoku, Japan, involved six major bridges carrying four traffic lanes and four railway tracks.

Summary

The transport of people and goods provides many different challenges for engineering and architecture. The basic starting point is not a single site on which a building has to be designed, but a series of movements, sometimes over a wide area, that has to be accommodated by the required infrastructure. Inevitably, transport systems and structures have become more complex in order to meet the increasingly onerous demands of the passenger and freight shipper. Compromise is frequently essential, but there are still opportunities for exciting design. In this book I hope to communicate some of this excitement to the reader, and to illustrate the state of the art and science of transport architecture and engineering at the beginning of the new millennium.

Left (figure 41): The engineer/architect Santiago Calatrava created bridges, stations and airport buildings in sculptural forms reminiscent of animal skeletons – as here on the Almeda Bridge in Valencia, Spain.

> Chapter 1
Airports

> Terminal 4, John F. Kennedy International Airport

New York, USA, 1994–2001

Idlewild airport was commenced in 1942, opened in 1948 and it was rededicated John F. Kennedy International Airport in 1963. Today it covers a site of about 2,000 hectares (5,000 acres), has nine terminals and four runways, handles about 31 million passengers a year and employs 37,000 people. The airport has 108 contact gates and a further 65 remote aircraft stands. Terminal 4, formerly the International Arrivals Building, was opened in 1958. In 1997 the Port Authority of New York and New Jersey (who operate the airport on behalf of New York City) signed a long-term lease with JFK International Air Terminal LLC (JFKIAT) to operate and redevelop the terminal, which had consistently come bottom in passenger surveys for many years. JFKIAT is a joint venture between airport operators Schiphol Amsterdam, New York developer LCOR and the financial house of Lehman Brothers. They adopted the following mission statement: 'To create New York's preferred international gateway by providing a pleasurable travel experience for all customers.'

The innovative public/private partnership between the Port Authority and JFKIAT, combined with well-thoughtout design, high-level and detailed phasing plans and a true collaboration between designers and construction managers, made it possible to trim five years off a conventional procurement, which would require ten years for design, tender, let contract and construction. The organizational model that tied each partner firm's commercial objectives to the success of the project as a whole can serve as an inspiring example to others looking to redevelop life-expired infrastructure throughout the world.

Previous page: The glazed wall at JFK Airport creates a welcoming atmosphere in the new terminal.

Opposite: The 72-metre (235-foot) clear-span roof of the check-in hall serves as a smoke reservoir, eliminating the code requirement for internal fire compartmentalization. The resulting unobstructed area provides flexibility to adapt the hall if user demands change over time.

Layout

The site covers 64 hectares (163 acres). The terminal building, which handles both international and domestic flights for over 40 airline operators, has 16 contact gates and 140,000 square metres (1.5 million square feet) of accommodation. It has been designed for expansion on a modular basis, and the completed programme will have 42 contact gates and 270,000 square metres (2.9 million square feet) of floorspace. Peak capacity will then be 7,450 arriving passengers and 5,450 departing passengers per hour. These high capacities are required to promote the hubbing operation of the principal tenant airline.

An elevated light rail station is integrated into the core of the terminal. The Port Authority is developing a Light Rail Transit (LRT) project that will link all terminals and connect the airport to the Long Island Rail Road and New York Subway systems. This is the first terminal project at JFK to bring a rail system into the building envelope.

The design and masterplanning of the terminal allowed for a doubling of capacity in later phases. From analysis undertaken during the planning stages it was decided that each phase of the terminal development should meet the needs of all parties for the design life of that particular phase. Projected flight schedules were analysed to determine the number of gates required and how many people would be arriving and departing and in the building at any one time, and road traffic forecasts for peak-usage conditions studied to determine kerb lengths and parking demand. From this detailed analysis a design balanced between terminal (check-in, security, immigration, baggage handling, retailing and lounge space), landside (LRT, kerb space, roads and parking) and airside (gates and taxiways) was achieved for each stage of the project, from commissioning of the first phase to completion.

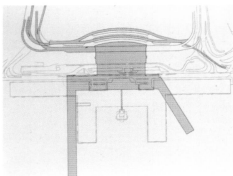

Above: The model of the new terminal shows how I. M. Pei's control tower will continue to dominate the aircraft on the apron.

Top right: The plan shows the location of the new building in relation to the old terminal.

Bottom right: The former terminal, before it was engulfed by the new building.

Opposite: The nine main stages of construction, showing the old buildings in yellow and new ones in red. Shown in blue are the temporary access routes through the 'Knuckles' which allowed the terminal to function throughout the reconstruction period.

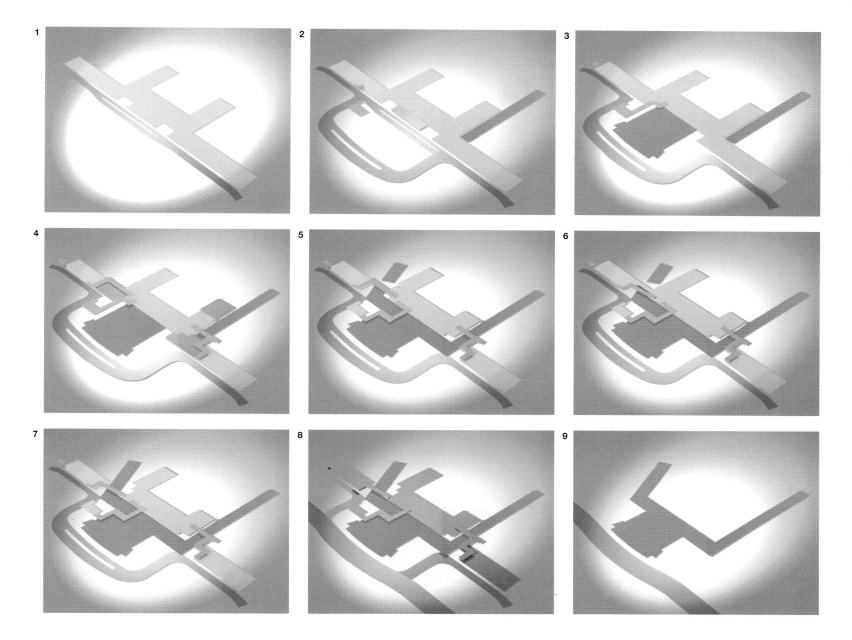

Programming and phasing of the works

The project was designed to minimize disruption to existing airport operations and allow the terminal to offer a complete service at all times. This required careful planning and co-ordination of all landside, airside and terminal functions.

The key to the phasing was the siting of the new terminal in relation to the old one. Given that they both required interface with landside and airside for the entire duration of construction, there was no option but to have overlapping programmes of work. Most of the new terminal is situated in front of and nested between the two wings of the old terminal. The new concourses were cut through the old building at what became known as the 'Knuckles'. While this was being done, passenger flows, airport operations and connections between all building utilities had to be maintained. This was achieved by a series of phased corridors and connectors, co-ordinated with the selective demolition of the old terminal and the release of zones for new construction. The phasing was further driven by the need to maintain a minimum of ten contact gates in operation at all times. The solution to this requirement was a series of nine airside stages. This led to the design of ten different pavement types in a co-ordinated patchwork on the airside to allow planes to start up, manoeuvre and park at different locations in each of the airside stages. By designing each part of the apron for its estimated usage, major savings were made on temporary works costs.

Opposite: The departures road is at the fifth level. The departures hall has a wave-form roof, and the control tower stands behind it.

Above: The main terminal, linking the access roads to the circulation building, straddles the route of the light rail transit.

Architecture

The most remarkable feature of the terminal is the ticketing hall, a glass and steel pavilion on top of a more conventional three-storey box. While the ticketing hall can be appreciated immediately for its architecture, its form was designed to meet the organizational, spatial and structural limitations of the phased construction and the desire for economy in both building and operation costs.

Together, the phasing requirements of the project and the footprint of the old 1950s terminal determined the shape of the new ticketing hall. By relocating the arrivals area to a temporary building it was possible to create a clear footprint – over 170 metres (560 feet) wide – between the two wings of the old terminal. This width proved to be sufficient to accommodate all of the ticketing area on the fourth floor and the baggage claim area on the first.

The most notable feature of the ticketing hall is its spectacular 72-metre (235-foot) clear-span roof. At 12 metres (40 feet) high, it is double vaulted and supported on exposed and sloping pipe-column bracing with large pinned connections. The need for such a high ceiling was driven by fire engineering studies of evacuation times which showed that, if the hall were to be filled with smoke, a roof-space smoke reservoir would allow people to exit the area safely. To facilitate this, clear exit routes were designated within the open-plan space so that people can always see outside and understand where to go. This avoided the need to introduce a series of fire-protected corridors and tunnels throughout the hall; it also allowed a derogation of the code requirement of a travel distance to a place of safety of no more than 46 metres (150 feet).

Structural and mechanical engineering

A concentric bracing system for the roof was developed since it is the most economical lateral system. The northern pipe braces form a pyramid to span the 18 metres (60 feet) over the LRT station and tracks below. This allowed the pyramid to be an isolated structure that could also provide an anchor for the pinned roof. The next line of sloped columns was 78 metres (235 feet) away, arranged in a V pattern to provide lateral stability in the east-west direction. A second line of V columns was located 27 metres (90 feet) further on, providing support for the second vault of the roof and symmetry with the long retail hall below. The double-vaulted trusses spanning from the pinned pyramids and V columns were designed with economy in mind. Pinned columns and closely co-ordinated architectural details allowed the roof structure to rotate and deflect, with ultimate movements of 203 millimetres (8 inches) south and 305 millimetres (12 inches) down under full loads. The scheme for the roof structure came in at a light 88 kilograms per square metre (18 pounds per square foot) for 78 metre (235 foot) and 27 metre (90 foot) double spans, compared with between 137 and 147 kilograms per square metre (28–30 pounds per square foot) for more conventional schemes. This gave savings of over US$7 million (£4.5 million) on the roof structure.

Once the high-volume space was specified, cooling it economically became the challenge. Air was introduced with high velocity nozzles at low level over the check-in counters and returned at an even lower level between the baggage belts behind the counters, thereby saving on the energy required to cool the entire 12-metre (40-foot) high volume. The system design, combined with an innovative IT system that controls building operations in response to airline operations, led to an impressive 15 per cent overall saving in energy for the building (when compared with other new energy-efficient airline terminals), calculated using an energy-life-cycle analysis. For the 25-year lease period of the project, this equates to over US$47 million (£30 million) in savings based on current energy prices.

Below: The pipe braces are formed into pyramids which span the light rail station and tracks below and form an anchor for the pinned roof.

Opposite: V-columns, supporting the northern end of the main vault of the roof, provide lateral stability.

Landside access infrastructure

In addition to the rebuilt terminal, landside access infrastructure has been renewed. In response to the Airport Authority's rearrangement of the road layout to provide direct access to the four quarters of the Central Terminal Area (CTA), new elevated roadways were required to match the revised external layout and to provide vertical separation between the arrivals and departures levels. The design also had to incorporate the guideway for the new LRT in the terminal area.

Hitherto, all public transport access to the airport has been by bus. Whilst the LRT system will provide access to the suburban networks, many international travellers now expect a high-speed train service to the city centre, thus avoiding traffic congestion. The Port Authority has plans for a One-Seat-Ride service, similar to the London Heathrow Express, to take passengers directly into Penn station in Manhattan. The dimensional specification of the LRT will allow a shared right of way with the proposed One-Seat Ride.

Below: The car parks and access roads were reconstructed to accommodate the increased capacity of the terminal. I. M. Pei's control tower is visible.

Opposite left: The management control systems in operation.

Opposite right: Passenger information is updated in real time directly from FAA air traffic control systems.

Information technology

Airport operations are highly dependent on communications systems. The systems required at Terminal 4 were classified as Building Management Systems (BMS) or Airport Information Management Systems (AIMS). The BMS were specified as commercial off-the-shelf (COTS) equipment, since proven COTS interfaces exist between the BMS and most other building IT systems, including lighting controls; elevator, escalator and moving walkway management systems; security; electric power management systems; sub-metering systems; and computerized maintenance management systems.

The issue of a conceptual design and the resulting tenders from systems integrators had not proved satisfactory for the AIMS. It was therefore decided to manage the process through the design team. JFKIAT thus retained ownership of all interfaces between systems. The AIMS was selected to serve at the heart of the airport management. A flexible approach was developed for two reasons: to enable individual applications to be upgraded in a modular way when improved technology becomes available or is required, and to accommodate the planned expansion of the terminal. Detailed specifications for each system were prepared because COTS systems are not readily available for these applications.

At an early stage it was decided that it was necessary for the terminal to be operated as a shared facility. This means that JFKIAT is responsible for the allocation and management of shared resources such as check-in desks, gates and baggage belts. This strategy required the development of detailed specifications for AIMS and the interfaces between AIMS and other airport systems. The AIMS required links with Official Airlines Guide (OAG) to obtain six-month, advanced seasonal schedules for planning, and with airline Departure Control Systems (DCS) to obtain revisions to flight schedules and updates to estimated times of arrivals and departures automatically. IT links allow the Federal Aviation Authority and other sources to provide real-time information during flights. The airport uses resources allocation tools for the automated allocation of terminal resources (gates, check-in desks, baggage carousels) and to update the flight-information display systems throughout the terminal automatically, and these too are linked into AIMS. The building management system relies on AIMS links to adjust the occupied/unoccupied schedule of each HVAC/lighting and security zone in order to minimize energy consumption and control security-staffing requirements, and the accounting system uses IT links for recording the resource usage of each airline and automatic billing. These IT systems therefore represent best current practice in airport terminal design.

Conclusion

Any architect working at JFK is conscious of the precedent set by the dramatic TWA Terminal (now Terminal 5) designed by Eero Saarinen and completed in 1962. Saarinen had the advantage of a relatively clear site and was able to create a new building. Today, a common requirement of airport terminal design is that existing terminals must be able to operate while the facility is upgraded to accommodate increased passenger capacity and the requirements of new-generation large aircraft. This constraint often results in a conservative solution and industrial-quality design. The JFK project shows that high-quality architecture and exciting engineering can be part of an economical and efficient solution to producing a building that will meet the requirements of the first decades of the 21st century. Physical, operational and cost constraints can lead to quality design solutions – and in Terminal 4 JFKIAT have shown just what can be done.

Above: The new terminal has 16 contact gates, but can be expanded on a modular basis to provide 42 gates as demand grows.

Opposite: The departures kerb provides direct access to the check-in hall.

> Terminal 2, Cologne/Bonn Airport

Cologne/Bonn, Germany, 1993–2000

The airport, situated about 15 kilometres (9 miles) south-east of the centre of Cologne, serves both Cologne and the former German capital city of Bonn, 20 kilometres (12 miles) to the south. Since the late 1960s the airport has been served by a concrete terminal building with a design capacity of 4.5 million passengers per annum (mppa). By 1992, with traffic up to 3.67 mppa, it was necessary to plan for expansion, and a competition was held to design a second terminal. The winner was Murphy/Jahn architects, working with Arup.

German-born Helmut Jahn joined the Chicago firm of C. F. Murphy Associates in 1967, and by 1983 had succeeded to role of president and CEO of the practice, now renamed Murphy/Jahn. Working closely with engineers and relishing the use of steel and glass, Jahn calls his architecture 'archi-neering'.

The architectural critic Michael J. Crosbie, in *Murphy/Jahn: Six Works* (Images Publishing, 2001), describes it thus:

'In archi-neering the architect and engineers genuinely collaborate, conjuring buildings that respond to the program and the technology of advanced materials and systems. But this is not a technocratic approach to architecture. Rather, it is using technology and science in the service of architecture that expresses a certain zeitgeist, and also meeting the needs of those who inhabit these structures. This is not technology for technology's sake, nor an aesthetic of "high-tech", but materials, equipment and methods of environmental control that seek to become invisible. Jahn notes that the ultimate building would be one that provides all the creature comforts virtually undetected – an architecture that serves and vanishes, simultaneously. Now collaborating with structural, mechanical, and environmental engineers, along with physicists and materials manufacturers, Jahn is creating buildings that push the limits of glazing technology, environmental systems and urban place-making.'

This approach resonated with the Arup engineers, who had recently set up offices in New York and Düsseldorf and were thus well placed to work in Germany with an American-based architect.

Opposite: The new terminal is constructed of prefabricated steel and glass components on a concrete substructure.

Terminal 2

As well as the new Terminal 2 building, which increases the capacity of the airport to 11 mppa, the project encompasses an underground inter-city express (ICE) and local (S-bahn) train station, two multi-storey car parks and a completely redesigned road-approach system.

The low-profile Terminal 2 follows the leg of the splayed-U configuration of Terminal 1. Whereas the existing concrete building is solid, the new building is constructed of prefabricated steel and glass components on an exposed concrete substructure, creating very light and transparent walls and roof. Terminal 1 has gates clustered around nodes on the outside of a U-shaped concourse, but in contrast Terminal 2 has a straight concourse providing a larger area of terminal space for each gate to accommodate the wide-bodied aircraft increasingly being used on short-haul and holiday charter flights. The arrangement of a linear concourse and frontal gates is clear and simple and one that both arriving and departing passengers can easily understand. Journeys to and from Parkhaus (car park) 2, the elevated roadway, the ICE station and the planes are pleasant, along light and well-signed routes, facilitating passenger movement both horizontally and vertically.

The new underground train station will link the airport to the European high-speed rail network via a 15-kilometre (9-mile) long loop railway line. This will result in two significant advantages: the airport is likely to become a central traffic hub for the west of Germany and thus an important and appropriate gateway for the nearby Rhine/Ruhr economic region, and many short-distance flights from and to other German or nearby European cities could soon be replaced by express rail journeys – a major improvement for the environment.

The new terminal building has a five-storey reinforced-concrete substructure. Level 1 has the train platforms, with the station concourse above. Level 3 is the arrivals hall, and the baggage-handling area on level 4 serves the departures hall on the top deck.

The departures hall

Architects often choose to make the departures hall the highlight of an airport terminal, because departing passengers spend more time in the airport than arriving ones. Helmut Jahn's design for this building is no exception. His design aims at lucidity and visibility: the space is surrounded by a glass façade and features a steel skylight with glass bands that allow daylight to enter, supported by slender, tree-shaped columns. To the north and south, atria penetrate the building over its entire height and thus connect the different levels visually. In addition, the terminal features light, steel framed floors with glass finish, steel stairs with glass treads and glass-clad elevators and glazed air bridges.

The roof structure of the new departures hall consists of a flat skylight supported by 22 tree-shaped columns located on a grid 30 metres by 30 metres (100 feet by 100 feet). In plan the roof comprises a 180 metre by 90 metre (600 feet by 300 feet) central zone, with two 60 metre by 30 metre (200 feet by 100 feet) wings. The skylight is a corrugated truss structure that enables two neighbouring trusses to share a top or bottom chord. The trusses consist of tubes and run continuously over the tree columns. At the roof edge the skylight is typically supported by the façade mullions. On the building's landside, however, the skylight cantilevers 15 metres (50 feet) beyond the façade line to provide a roof for the set-down area in front of the building.

The tree-column solution to support the skylight is not only visually attractive, but it allows a relatively large column spacing while keeping the skylight structure shallow and the roof shape flat. The trunk of a tree column consists of four vertical tubes with a centre-to-centre spacing of 2.5 metres (8 feet 4 inches). The typical trunk measures 2.6 metres (8 feet 6 inches) in height, except for those along the landside façade which are supported at a lower level and are 13 metres (43 feet) high. The maximum spread of the branches at the top of the tree is 32 metres (105 feet). Typically, the clear height between the finished floor of the departures hall and the bottom chords of the skylight is around 9 metres (30 feet). Pin connections and 50-millimetre (2-inch) thick node plates join the top branches of the columns to the chords.

The slender, tapered branches of the tree columns are perhaps the most significant architectural feature of the steel roof structure. The tapered portions of the branches are seamless and were manufactured using a method whereby solid, square steel blocks are first pierced with mandrels, then the excess material forged and removed. This process is repeated until the specified diameters and wall thicknesses are achieved. The fork-and-eye end pieces of the branches are castings that continue the same taper as the forged tubes. This results in very elegant connection details. The ends of the diagonal tubes in the skylight trusses were cut in curved shapes with fully automatic flame-cutting machines. The tube-to-tube connections in the tree trunks were hand-cut with computer-generated templates. All tubes in the roof structure are seamless.

The roof itself consists of panels, so-called cells, which are placed onto the folded plate with simple bolted connections and waterproof joint seals. These cells are designed to fulfil various functions: light transmission, weather protection, exterior heat absorption, interior heat absorption, acoustic dampening and absorption, and smoke ventilation. The aim was to create a skin with self-adapting qualities by combining different types of cells, and the resulting design became the prototype for a series of buildings. In effect, the roof or façade is no longer a cladding with constant properties but the technical equivalent of biological skin.

The façade is a lightweight cable-supported steel-and-glass structure, with insulating glass panels held by 'spiders' at their joints. Similar advanced technologies are applied to the glass railings, elevators, fixed and moveable jet bridges and glass floors and stairs.

Conditioned air is supplied to the departures hall via air columns integrated into the steel trees supporting the roof. Return air is drawn down from the ceiling zone through the air columns and routed back to the main air-handling stations located below the baggage claim hall. Fan-coil units recessed along the façades provide supplemental heating and cooling. Within the volume of the departures hall only the lower 3 metres (10 feet) will be heated or cooled to provide a comfort zone for passengers. This stratification of temperature creates an economically conditioned layer at pedestrian levels and an effective thermal buffer zone to the outside at the upper levels.

Above: The new terminal is on the left of Terminal 1, which was designed for 4.5 mppa but was handling 6.4 mppa by the time the new terminal was completed. The high-speed railway station is beneath the new terminal.

Opposite: The 'Christmas Tree' columns, reminiscent of those at Foster's London Stansted terminal, provide large clear spans as well as accommodating mechanical and electrical services.

Car parks and railway station

The car parks, in which Arup had no active involvement, were kept as low as possible in order to preserve the views across the airport landscape to the existing Terminal 1 and new Terminal 2. On both of the multi-storey car parks the same system components were used. Their structure comprises 65-metre (210-foot) wide units separated by a 10-metre (33-foot) wide light court. Highly detailed steel structures clad the basic rough-steel construction. They include the stainless steel-mesh façade cladding, the vine walls in the light courts and railings of stainless steel cables. The elevator/stair towers are also steel structures, with cantilevering platforms of stainless steel planks in front of the point-fixed glazing of the elevator towers. The glass elevator cabs ride in open-air shafts and are designed to function in extreme weather conditions.

Parkhaus 2, the first building completed in the airport expansion, lies on a trapezoidal site bounded by the existing elevated roadway, the new Terminal 2, the future train station and the new road network. It has 5,800 parking spaces on six levels, linked by two cylindrical ramp structures. The upper level has an additional entrance/exit directly to and from the elevated departures roadway in front of Terminal 2 and is intended for short-term parking. The lowest level is dedicated to rental-car return and pick-up. Airport visitors have direct access to the new Terminal 2 via levels 2, 4 and 6. The intervening levels are linked by the two open-air elevator/stair cores, ensuring easy visitor orientation and quick access to the terminal. The eight-storey Parkhaus 3 has 4,500 long-term parking spaces.

The four-track train station is covered by a slightly arched, cable-supported glass roof, which is 200 metres (660 feet) long, and which projects above ground level to provide natural daylight. The station box will be completed when the new high-speed railway line from Cologne to Frankfurt is constructed.

Conclusion

At the start or end of a journey, Terminal 2 and its associated structures are an introduction to a technology-based region, of which the airport itself and the cities of Cologne and Bonn are a part. Airports are frequently the focus of civic pride, and there is certainly justification for this here.

The Terminal 2 project is an excellent example of a smooth design process followed by close and productive collaboration between designer and contractor during the fabrication and erection phase. The structural and environmental engineers were innovative in meeting the challenges posed by the architectural concept, and the application of modern steel-fabrication methods helped to realize a structure that both satisfies the client and represents an eye-catching product of archi-neering.

Opposite: Clear spaces create an uncluttered check-in hall with natural lighting.

Above: The new car park incorporates a series of easy-access decks.

Below: The terminal departures kerb is over the arrivals kerb, with the car park providing multi-level access.

> New Terminal, Lester B. Pearson International Airport

Toronto, Canada, 1997–2004

Lester Pearson Airport is the busiest airport in Canada, handling around 28.9 million passengers in 2000. It is located 25 kilometres (16 miles) north-west of the Toronto Central Business District, and serves the southern Ontario region as well as being a hub airport for Air Canada. The airport's three terminals are adjacent to each other on the eastern side. By 1995 it had become evident that these buildings were clearly unsatisfactory and could not readily be extended to provide the capacity needed when the airport expanded to six runways. Traffic is forecast to grow to 39.9 million passengers per annum (mppa) by 2010 and 49.7 mppa by 2020.

Terminal 1, opened in 1964, comprises a circular building with passenger processing at the centre, gate lounges in a circular ring serving 23 air bridges, and parking above the terminal area. At the time, the masterplan for a series of such circular terminals, accessed by a road tunnel beneath the apron, was at the leading edge of airport design. The concept was not followed through, however, and a similar design was not adopted for Terminals 2 and 3.

Terminal 1 no longer meets modern requirements and is to be demolished. Terminal 2, a linear two-level structure, is used mainly by Air Canada and its associates. Opened in 1972, it was originally planned as a temporary building, but there have been subsequent extensions, including a satellite used by transborder passengers. The site and building geometry of Terminal 2 limit further expansion opportunities and the ability to provide acceptable levels of service to a larger number of passengers. Terminal 3, opened in 1991, has a central processing area with two piers serving 29 gates in a horseshoe configuration. It is in good condition and will be subject to some expansion, complementary to the new terminal which will replace Terminals 1 and 2.

The Greater Toronto Airport Authority (GTAA) assumed responsibility for the airport in 1996. With the assistance of ArupNAPA, the airport planning division of Arup, the authority has developed a comprehensive redevelopment plan to enable it to keep pace with the growing travel needs of Toronto and the southern Ontario region. The Airport Development Program (ADP) is a 10-year, C$4.4 billion plan covering four major projects: terminal development, airside development, infield cargo development and utilities and airport support. The first three stages of this development programme are due to be completed by 2007, with long-term development, including the fourth stage of the new terminal building, to be phased subsequently as demand dictates.

Opposite: The curved terminal building is divided into segments, which each serve a pier containing the holdrooms. This allows for modular expansion of the terminal and the replacement of redundant buildings.

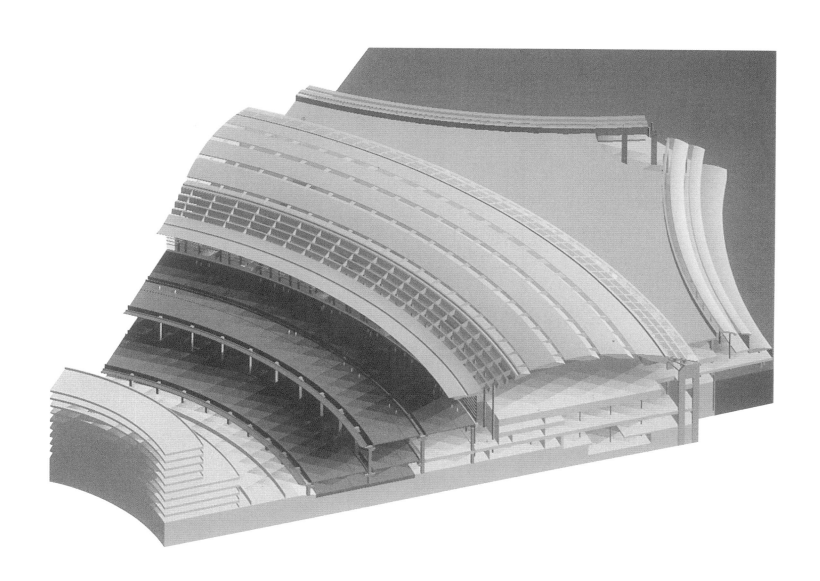

Design brief for the new terminal

The masterplan provided for a four-concourse terminal of 420,000 square metres (4½ million square feet) to replace the existing Terminals 1 and 2 in three development stages. SOM, in association with Adamson Associates and Moshe Safdie & Associates, were appointed in 1997 to design the building. The redevelopment had to meet three challenges that have become a common feature of the design of airport terminals. Firstly, it had to ensure that existing facilities could be kept operational, with the terminal built over and around them. Secondly, it had to be able to accommodate changes in the brief as circumstances altered during the design and construction period: such changes are not unusual in the dynamic environment of commercial aviation. Finally, it required fast-track construction to reduce the period of disruption and bring the new facilities into use as soon as possible. For this, close co-operation between client, the design team and the contractor was essential.

A particular complexity at Toronto was the requirement to keep separate three classes of passenger – domestic, international and transborder (US-bound). The brief for the new terminal required four passenger flows to be segregated – domestic (arrivals and departures), international departures, transborder departures and international/transborder arrivals. The design challenge was to find a way to process these flows in an efficient and user-friendly way without creating a rabbit-warren and compromising the clarity of the architecture. Because the peaks for different sectors occur at different times, maximizing gate flexibility was a further challenge.

GTAA wanted a common-use terminal, with a flexible assignment of facilities such as check-in counters, gates and baggage-handling equipment, but this was not welcomed by Air Canada, the dominant user, who wanted their own exclusive areas. The design, therefore, had to meet Air Canada's operational requirements wherever possible while maintaining the common-use flexibility.

Below: The new building in the foreground replaces Terminals 1 and 2 in Stages 1 and 2 respectively. Terminal 3, in the background with the Terminal 3 satellite beyond, was completed in 1991.

Opposite: The main hall of the terminal has a roof span of 70 metres (230 feet) across a curved space 350 metres (1,150 feet) long.

The architectural solution

The key to the terminal planning was the layering of the three sectors, domestic, transborder and international, in a way that would minimize connection times, particularly between domestic and transborder flights, the biggest growth market for Air Canada. Transborder gates and the US pre-clearance facilities were therefore strategically placed in the middle of the terminal, between the domestic and international boarding areas. The ability to transfer passengers between international and transborder flights without passing through Canadian customs and immigration offers the airport the opportunity to strengthen its position as a competitive international gateway to the USA.

The plan proposes an arching terminal processor with four main piers serving up to 105 bridged gates. In the long term, the processor can extend east to replace Terminal 3, adding two further piers and up to 50 additional gates. Primary access to the terminal area is moved to a new highway system leading from the north, and there will be an eight-level parking structure with 12,600 spaces.

Apron planning provides for every gate to be served by two taxiways. Stand depths are set to allow Boeing 747 parking, plus ample provision for the next generation of large aircraft. Both head- and tail-of-stand roads are also proposed. The resulting airside configuration dramatically improves both aircraft movement and operational flexibility.

The plan was conceived with a three-level curb and a five-level terminal. The lowest curb, one level below the apron, provides service access to the building and also accommodates charter-group buses and facilities for airport staff. The apron-level arrivals curb of the processor corresponds to a large arrivals hall wrapping the ground-side face of the terminal, with a two-storey baggage claim hall beyond. Baggage make-up facilities are also at this level within each pier. Above this the mezzanine level of the processor has bridge links to the parking structure. This level corresponds to the gate-lounge level of the piers and the primary inspection line of the Canadian Inspection Services, which overlooks the international baggage claim hall. The departures curb and check-in hall are on a level above this, with passenger-circulation mezzanines in the piers. The check-in hall is a remarkable space, with a 70-metre (230-foot) clear-span arching roof enclosing a space some 350 metres (1,150 feet) long. A fifth level is provided for airline lounges at the nodes where the piers meet the processor.

The gates are organized to provide maximum flexibility. Domestic services are located on the most easterly main pier of three in the initial development, with commuter aircraft served from a small lateral pier. International gates are primarily located on the head of the second central pier, with some additional capacity on the head of the third pier. Transborder gates are accommodated between these groups, and there is considerable ability to shift between transborder and each of the other two sectors on an hourly and seasonal basis and over longer periods.

When the first three phases are completed, the terminal will have over 420,000 square metres (4$^1/_2$ million square feet) of space. In comparison with existing terminals, this new facility represents a quantum leap in the quality, flexibility and efficiency of facilities. The flexibility of the design has already been put to the test – significant upheaval took place in the airline industry and major changes in company ownership took place during the design and construction.

Staging

Terminal 1 and part of Terminal 2 will be demolished on completion of Stage 1 of the new terminal in 2003. Stage 2 will permit the demolition of the rest of Terminal 2. The Stage 1 works are limited by the need to keep the existing terminals fully operational during construction. Stage 1 includes most of the central processor, where check-in, security screening, immigration, baggage reclaim and customs inspection take place. All passenger processing from Terminals 1 and 2 will be consolidated in the new terminal by 2005.

The phased construction of the airside concourses and aprons means that the gate processing capacity will not match the processing capacity until Stage 3 is complete in 2008. Until that time, up to 30 per cent of passengers will be processed in the new terminal and bussed to remote holdrooms.

Six bus lounges capable of taking all classes of passenger are planned close to the main terminal facilities so as to minimize walking distances.

11 stages are required, to be implemented over 13 years, which indicate the complexity of the operation and the value of a long-term planning framework.

Airport terminals differ from any other class of building in the demands they make on the skills of architects, engineers and project managers. The complexities of staging construction in and around a working building are massively complicated by the separation of passenger flows and other security arrangements that are required at an international airport. Demands are constantly changing; the increased security measures following the events of 11 September 2001, for instance, are just one example of the modifications that have taken place during the evolution of this project.

Below: The simple lines of the exterior of the terminal disguise the complexity of the internal arrangements.

Opposite: This view of the ticketing hall at the upper level shows how the roof structure provides a welcoming character to the building.

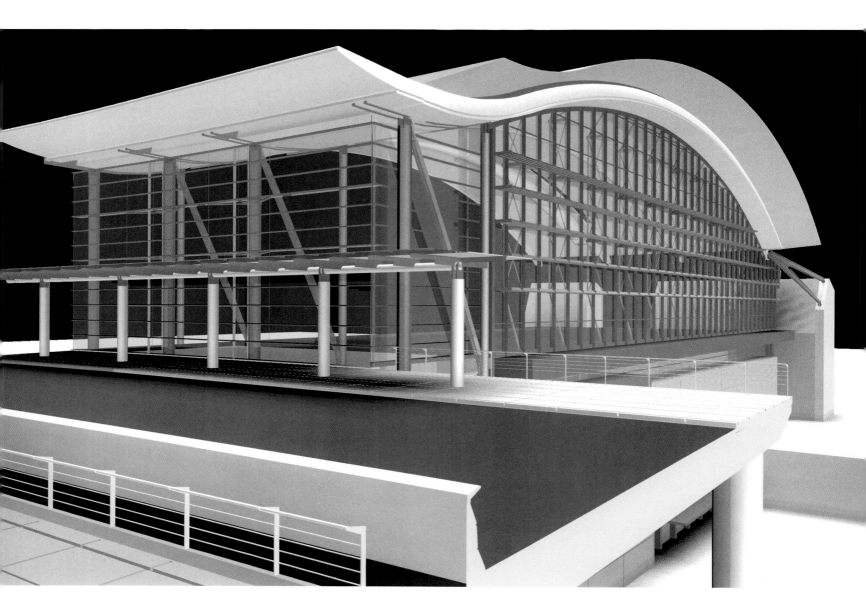

Industry changes

When Air Canada acquired Canadian Airlines in April 2000, there was an immediate change as the company required co-location of services that were previously separated. Forecasts were also revised, with domestic traffic predicted to be higher and international traffic reduced. The built-in flexibility of the design has accommodated these changes without the need for serious re-planning.

Fast tracking

GTAA set a tough construction and staging schedule. But because the design process was synchronized with the timing of construction, parts of the building had to be designed, tendered and built in a continuous fashion. For example, while steel for the central processor was going up on site, parts of the facility were still at the schematic design stage. Early decisions therefore fixed the envelope within which later design had to fit.

This type of working requires close co-ordination between client, design team and contractors to ensure that decisions are made in a timely fashion and that programme changes do not result in costly design changes.

Conclusion

Lester Pearson Airport represents another landmark terminal design for the Arup engineering team, whose track record includes the recently completed Terminal 4 at JFK (see pages 48–59), Chek Lap Kok in Hong Kong (see pages 72–87), Cologne/Bonn in Germany (see pages 60–65), Kansai in Japan, Stansted in the UK and the recently approved Terminal 5 project at London Heathrow. Like each of these projects, Pearson's new terminal carefully integrates the structure, building systems and architectural concept to achieve a large and dramatic but user-friendly terminal that will make arrivals and departures from this airport a pleasure. The project also anticipates expansion and change, with future phases of development carefully considered in the design.

When the new terminal opens in late 2003, it seems certain to set another benchmark in terminal design, much as the original Terminal 1 did in 1964. But this time, the masterplan provides a much firmer footing for the continued growth and development of Canada's premier airport.

> Hong Kong International Airport, Chek Lap Kok

Hong Kong, China, 1991–1998

The original Hong Kong Airport at Kai Tak had long been considered unsatisfactory for such an important and busy city. Landing aircraft had to bank several times to avoid hills and buildings on the approach over Kowloon, and the shape of the airfield, a reclaimed finger of land, put a limit on its capacity. In 1989 the Hong Kong Government Port and Airport Development Study (PADS) named the island of Chek Lap Kok off the north coast of Lantau Island as the site for a new airport to replace Kai Tak. The site was chosen for its clear airspace, as take-off and landing is over water. It also has expansion capability and is close to the urban areas of Kowloon and Hong Kong Island.

The total project was a massive feat of civil engineering. First, land for the airport site had to be reclaimed, increasing the original 302 hectares (746 acres) of the island of Chek Lap Kok to 1,248 hectares (3,084 acres). The rail and road links required a new cross-harbour tunnel from Hong Kong Island to reclaimed land on the western side of the Kowloon peninsula, as well as bridges to Lantau via the island of Tsing Yi. The railway comprises two lines: the Airport Express Line (AEL), providing a rapid link from Hong Kong and Kowloon to the airport; and the Tung Chung Line (TCL), a domestic line between Hong Kong, Kowloon and the new town of Tung Chung on Lantau. Most of the route is common between the two lines and some bridge and tunnel sections share tracks.

The airport masterplan, finalized in 1991, defined the layout of the airport platform and its two runways, the location and general arrangement of the Terminal Building, and the zones for cargo handling, aircraft maintenance, fuel storage, catering and other commercial activities. Originally the terminal had 38 gates served by bridges and all capable of accommodating wide-bodied aircraft. The Ground Transportation Centre (GTC) is adjacent to the terminal, which has already been expanded to 48 gates. The GTC contains the AEL station on two levels; arrival and departure areas for buses, coaches, hotel vehicles, taxis and private cars; parking; and an area for baggage handling and transfer of baggage checked in at Hong Kong or Kowloon.

Other buildings on the site provide for cargo, catering and all the other support facilities required by a major modern airport.

Current traffic through the airport is 33 million passengers per annum (mppa), similar to the numbers passing through New York's JFK, London Gatwick or Amsterdam Schiphol.

Opposite: The barrel vaulting of the canopies carries the main structural theme of the terminal over the departures kerb of the Ground Transportation Centre, welcoming the traveller to the airport.

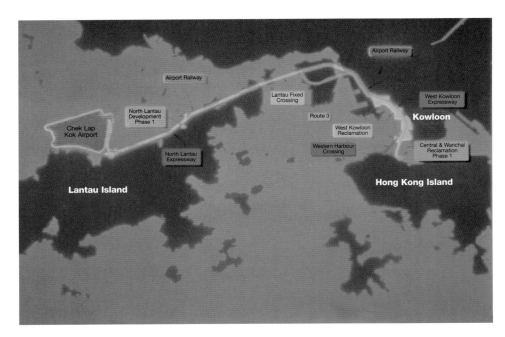

Left: A series of bridges and tunnels connects the airport site to the commercial districts of Kowloon and Hong Kong Island by road and rail.

Below: The rocky islet of Chek Lap Kok was flattened and, with the help of a massive dredging operation, a new island was created for the airport.

Opposite: The distinctive barrel vaulting of the terminal roof provides spans of 36 metres (120 feet), and with the simple columns provides large, clear spaces.

The Terminal Building

The detailed design of the Terminal Building was the subject of competitive tender. The masterplan assumed 35 mppa initially, ultimately rising to 87 mppa. It also defined the number of aircraft stands, the principal space requirements and how the building was to be phased.

The key components of the terminal are its links to the GTC, the processing building, the concourses and the passenger transit system within the building. The basic concept of separation of activities is similar to that in operation at London Stansted Airport, also a collaboration between Foster and Partners and Arup. Baggage handling and the mechanical engineering plant are in the basement. There are then two main passenger levels, with the departures level above the Arrivals Hall. Further plant, airline offices and retail outlets are placed in mezzanines between basement and arrivals, between arrivals and departures, and above departures.

The five main elements of the terminal, from east to west, are:

– The Processing Building. This contains check-in, security, baggage handling, baggage reclaim, immigration and customs, and the Arrivals Hall. The North and South Concourses lead directly from the Processing Building to gates north and south of the core of the building. The East Hall is the main shopping area and also provides access to the people mover in the basement which takes passengers to and from the West Hall for access to the diagonal concourses. The roof is at its highest (22 metres/72 feet) over the departures level and falls over the East Hall to provide the minimum comfortable height of 4.9 metres (16 feet) and minimize both cladding costs and the air-conditioning load.

– The Central Concourse. This gives direct access to gates on both sides and has travelators at arrivals and departures levels. Light wells between the travelators in the departures level let in natural light to the arrivals level below. The automated people mover (APM) is below ground level with platforms about 6 metres (20 feet) below the Arrivals Hall. This permits movement and storage of vehicles and leaves space for plant rooms on the apron level beneath the concourse.

– The West Hall. This forms the junction with the diagonal concourses and provides access to and from the station at the western end of the APM at basement level.

– The two diagonal concourses. These are similar in layout to the central concourse, without the basement people mover.

Opposite: The view through the glazing overlooks Lantau.

Above: Bridge links lead to the check-in desks on the departures level, which is above the arrivals concourse.

There is a strong emphasis on modular construction for the roof areas, which comprise a series of vaults with 36-metre (118-foot) spans. The construction of the roof is based on a diagonal steel grid which allows the same design theme to be continued into the diagonal concourses at the western end of the terminal. With the vertical separation of arriving and departing passengers, the grid provides adequate width for the concourses in a single span.

The modular nature of the roof structure offered possibilities for prefabrication. The successful tenderer originally proposed to construct elements up to 36 metres by 36 metres (118 feet by 118 feet) – complete with finishes, services and the supporting structure – and deliver them to the site on barges using the large-scale modularization techniques developed in the offshore industry. In the event, however, the elements were cut to length and despatched from the works in Singapore in standard-size containers. An on-site fabrication yard was set up, which at the peak of the process was

employing 700 workers and processing 2,500 tonnes (2,460 tons) of steelwork a month. Sub-assemblies of 18 metres by 6 metres (59 feet by 20 feet) were constructed and then linked into assemblies up to 54 metres by 36 metres (177 feet by 118 feet). These were lifted into the paint shop with a 500-tonne (492-ton) capacity mobile crane and finally transported to the site using four self-propelled multi-wheel transporters.

The ties to the roof vaults are raised off the horizontal. This was primarily for architectural reasons but also to facilitate fixing the required curvature, which had to take into consideration the prevailing temperature. Calibrating each tie at the time of erection ensured they would all remain parallel after the imposition of dead loads and the effects of differential thermal expansion.

The Ground Transportation Centre

The Ground Transportation Centre (GTC) is the hub for all land transport to and from the airport. The original airport masterplan proposed a station within the Terminal Building envelope, but it was decided at an early stage to form a separate transport interchange. As a result, the GTC will be able to serve a future second terminal building.

The road system was designed to be free flowing up to 2010, with expansion capability to 2040. Principal traffic movements are grade separated and vehicles can circulate between the major car parks, drop-off kerbs and other facilities.

The GTC has five levels. The lowest level is underground and has three structures:

1 An area for processing baggage checked in at Hong Kong or Kowloon stations on the AEL, linked by a tunnel to the Terminal Building.

2 A tunnel and maintenance facility for the APM system running under the GTC and out to the aircraft gates.

3 A second tunnel allowing taxis to exit the GTC without conflicting with pedestrians.

The ground level is occupied by road traffic with separate, covered pick-up areas for buses, coaches, hotel vehicles and taxis. Private car pick-up facilities are in the car parks at each end of the Terminal Building.

The third and fourth levels form the arrival and departure platforms of the AEL. Because the two platforms are separated vertically, arriving passengers have a level walk from the baggage hall to the trains and departing passengers can walk on the level from the trains to check-in. These connections are via bridges across a 230-metre (755-foot) long atrium which runs along the length of the GTC on the side of the terminal building. The atrium is the main architectural feature of the centre. It is formed from curved vertical members spanning between the ground and the roof and 2 metre by 3 metre (6 foot by 9 foot) glass panels.

Above: The atrium runs the full length of the Ground Transportation Centre, separating it from the Terminal Building.

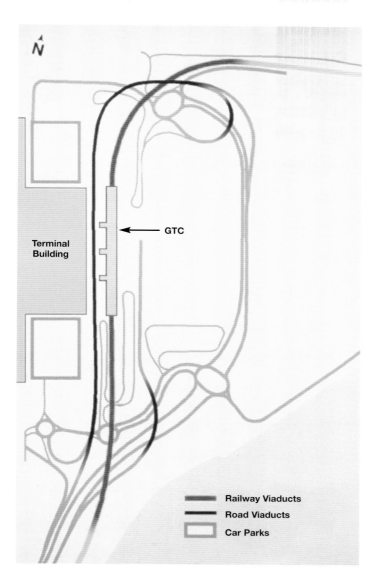

Terminal
Building

GTC

N

Railway Viaducts
Road Viaducts
Car Parks

At the fifth and highest level, the departures kerb is above the Arrivals Hall, with a median strip in the forecourt providing 650 metres (2,130 feet) of kerb along the Terminal Building façade for the setting-down of passengers.

The rail station is mainly a beam-and-slab concrete structure, with an exposed, cast-in-situ concrete roof enclosing large ducts for train and station ventilation. As with all stations on the airport railway system, platforms here have screens with doors that line up with train doors. As well as being a safety measure these screens retain conditioned air in the platform areas.

The design of the GTC reflects the particular features of transport in Hong Kong. These are high rail use and a large number of courtesy buses and limousines providing direct links to hotels. Car rental is insignificant, because Hong Kong's crowded roads and limited parking spaces make chauffeured travel a more attractive option. The raising of the rail lines to provide convenient and level transfers in and out of the terminal is a particularly attractive feature. The elevated lines also reduce the demands on vertical circulation within the building.

Left: A complex system of ramps and viaducts was necessary to separate the various road-vehicle flows and provide access to the buildings for both passengers and deliveries.

Below: Arriving and departing Airport Express trains are on different levels, providing level access from platform to Departures Hall and from Arrivals Hall to platform.

HACTL SuperTerminal 1

Most freight at Hong Kong International Airport (HKIA) is handled by Hong Kong Air Cargo Terminals Ltd (HACTL), who transferred their activities from Kai Tak. The planning of the new SuperTerminal drew on the experience of achieving increased capacity in the smaller spaces at Kai Tak. The buildings at Kai Tak had been operationally satisfactory but were essentially functional warehouses. At the new airport, HACTL wanted much more – flagship premises providing a high-quality working environment which related adequately to both the internal systems employed and the Terminal Building. The resulting structure, known as SuperTerminal 1 (ST1), has a quality of design and finishes seen more often in office buildings than a cargo centre.

ST1 also has a remarkable range of staff facilities. There is a fully equipped 1,500-square-metre (16,146-square-foot) sports centre with three squash courts; a swimming pool; badminton, tennis and basketball courts; a jogging track through the landscaped roof garden; and lockers and showers for all staff. On the southern edge of the building there is an executive dining terrace overlooking the roof garden. At the heart of the building is a triple-height glazed atrium containing exhibition space and a staff common room.

The company essentially processes goods from one mode of transport to another. As Anthony Charter, HACTL's former Managing Director,

succinctly put it: 'We take small boxes (for export) and put these into larger ones, and the reverse for import!'

The two functions of the operation require separate handling facilities and are accommodated in two buildings, The Express Centre and The Terminal, alongside each other on the south side of the airport. The Express Centre handles parcels and express cargo for the main courier firms, who occupy their own dedicated space and manage their own operations. These firms provide overnight services and often use their own scheduled aircraft to continental hubs, with long hauls between continental hubs being made by larger aircraft. The Terminal handles bulkier and less time-sensitive freight, much of which is transported as belly cargo in passenger aircraft. The combined capacity of the two buildings is 2.6 million tonnes (2.5 million tons) per annum. At the time ST1 was designed, this figure was about double the throughput of Heathrow, the largest airport in the world in terms of international passenger numbers.

The Express Centre is 200 metres by 90 metres (650 feet by 300 feet) on two levels and can handle 200,000 tonnes (197,000 tons) of cargo a year. Its primary role is to give the express cargo and courier operators their own offices and despatching and sorting facilities. The curved roof structure emphasizes the division of the facility into seven bays on each level. The northern bookend bay houses HACTL's own ramp-maintenance facility, associated offices, training rooms, rest areas and changing rooms. The southern bookend bay has an automated 6.1-metre (20-foot) container for storage, and a handling system for large items such as luxury cars and racehorses. There is also a strongroom area with four airside and four landside armoured truck docks, and a central security vault for consignments such as cash, diamonds and gold bullion.

A typical bay comprises an operational express warehouse floor some 36 metres by 45 metres (118 feet by 148 feet) in extent with a service core shared with the adjacent bay. Goods lifts in these cores link the operational floors to the central Customs and Excise Hall on the mezzanine above the ground floor. The operations carried out here are straightforward but labour intensive. Two-level operation is possible with the aid of two large hydraulic lifts capable of carrying 20-metre (66-foot) long dolly trains.

The Terminal is 200 metres by 290 metres (656 feet by 951 feet) in plan and has 240,000 square metres (2.6 million square feet) of operational, office and ancillary floorspace over six levels. Fully computer controlled, the building acts as a giant conveyor along which robot stacker cranes lift, pigeonhole and store cargo, enabling it to be unpacked and processed through Customs and Excise before being loaded onto lorries for delivery. At the heart of the Terminal are the two central Bin Storage Systems (BSSs) containing bulk cargo from, or destined for, road vehicles. The handling spaces either side include the

Above: Large clear-span roofs provide flexibility for the movement of trucks, fork-lifts and automated pallet transporters.

Opposite: Goods are stored on many levels, with automated equipment taking cargo into and out of the storage locations.

customs areas and the facilities for loading and unloading air cargo containers. A total of 130 automatic bin carriers transfer import cargo into the BSSs, link each BSS and transfer cargo between parts of the operational levels to the central Customs and Excise Hall on level 1 before transfer to the road-vehicle loading docks, also on level 1. The process is reversed for export cargo.

On the east and west façades are structures for the Container Storage Systems (CSSs) where air transport containers awaiting despatch or unloading are stored.

The arrangement of the CSSs is a key design feature. Their separation from the main building by a 16-metre (52-foot) roadway resulted in six building faces having airside vehicle access. The number of airside access points determines the capacity of the terminal and this created nearly 2 kilometres (1¼ miles) of frontage for loading and unloading dollies. Separating the CSS from the main building also meant they could be constructed, tested and commissioned before the main building was complete. This was an important factor in a tight construction programme.

At the northern end of the Terminal at ground- and first-floor levels is a perishable goods handling area. This is sited adjacent to the airport apron to minimize the distance between aircraft and the truck docks. Automated cargo hoists transfer goods between floors and to the 3,600 square metres (38,750 square feet) of large-scale industrial cold rooms and freezers on the second floor.

HACTL's central computer system tracks any item of cargo from the time it leaves its country of origin until it is collected. The office suites on the north and south elevations provide 12,000 square metres (129,000 square feet) of space for HACTL staff, airline representatives, government departments, banks and recreational facilities.

The structure of the Terminal is relatively simple. The planning grid, dictated by cargo-handling and truck-docking requirements, is 10.5 metres by 13.5 metres (35 feet by 44 feet), enlarged in two areas to 21 metres by 13.5 metres (69 feet by 44 feet) for road-traffic manoeuvring. The frame and decks are of reinforced concrete, detailed to allow for fork lift operation.

The most attractive parts of the structure are the roofs over the CSS roadways, the BSSs and the exhibition area. The architect's concept for these was based on a delicate butterfly-wing truss. The freestanding structures are separated from the rest of the Terminal by roadways crossed by bridges at the upper levels. The CSS roadway trusses comprise a single bottom chord and two top chords, which together act as a Vierendeel, with buckling of the top chord restrained by three horizontal purlins between each butterfly wing. The trusses were used as props to brace the CSS racking structures, which otherwise would not be able to resist typhoons without unacceptable distress to the external glazed façades.

For fire protection, because applied coatings were inadequate, a water-filled tubular system was developed. The water had to be circulating in order to reach the required thermal capacity, and this was achieved by providing sprinklers within the tubular structure itself and connecting the whole system to the fire service drencher system.

Lufthansa Catering Facility

Lufthansa Sky Chefs (LSG), the largest airline caterer in the world, was one of two such firms operating at Kai Tak. At the new airport they required a facility which could accommodate the growth in capacity forecast and serve their 25 airline customers.

Flight kitchens are complex facilities. Each airline has its own style and menus. Catering units have to look after not just the food but all the other items needed on an aircraft, including blankets, cutlery, duty-free goods, traveller kits and so on. Separate stores are therefore needed for each airline, as well as bonded stores and customs facilities.

The main elements of the building include a high-bay warehouse, chillers and freezers, and kitchen and dishwashing equipment. The high-bay store is 50 metres (164 feet) long, 25 metres (82 feet) high and 13 metres (43 feet) wide. It can store 2,400 loaded pallets weighing up to 1 tonne (0.98 ton) each. The goods housed include canned fruit and vegetables, alcohol, cutlery and blankets. Each pallet is bar-coded and the computer control system maintains an inventory of its contents and location and the expiry date of the product.

The transfer of business from Kai Tak to the new airport was a major operation. In total, 105 containers of goods and equipment were moved, 25 of them at night between the last serviced arrival at Kai Tak and the first departure from the new airport. The new warehouse had been pre-stocked, and this enabled a trial run to take place with the warehouse equipment. The offices were moved in advance of the changeover, with 61 containers of operational and office equipment being transferred before the closure of Kai Tak. Service carts from flights that arrived after the cleaning lines had been closed down and were transferred to the new site by road, either in containers or in LSC high-lift trucks. Overall the operation ran smoothly, with no break in service to the airlines.

Above: Lufthansa serves 25 different airlines, all with their own menus and branded accessories, and stores duty-free goods for sale on board as well as airline meals. The warehouse can store 2,400 loaded pallets.

Opposite top: The move from the old airport at Kai Tak involved transporting 105 containers of equipment, 25 of them overnight. The kitchens and staff had to be ready to start full operation on the new site only a few hours after serving the last flight from Kai Tak.

Conclusion

Building a completely new airport is a massive undertaking, involving far more than the provision of runways and a terminal. Surface access, accommodation for cargo, maintenance facilities and many other airport activities must all be in place before the first passengers are carried. When the new airport is to replace another which must close at the same time, the logistical challenge is enormous. At Chek Lap Kok the site as well as the buildings had to be created. Arup was involved in many of the airport related projects, all of which had to be completed simultaneously for opening day. Large teams were therefore mobilized which used both local and international resources of the firm, in order to ensure that all the new buildings were available that day. The completion of the airport to a tight timescale has not been at the expense of dramatic architecture: attractive buildings have been created for the use of passengers, and workers in various activities have been provided with uplifting and comfortable buildings from which to serve the public.

Below: Exterior view of the catering building.

> Chapter 2
Railways

> Hong Kong Airport Railway

Hong Kong, China, 1992–1998

The new airport at Chek Lap Kok is served by a new mass transit railway (MTR), the Airport Express Line (AEL), providing a rapid connection between the business areas of Hong Kong and the airport. A second railway, the Tung Chung Line (TCL), along much of the same route and sharing some stations and track, provides an urban metro service to new development areas created on reclaimed land as part of the airport infrastructure projects. The railways are designed to provide easy interchanges with other modes of transport, and airport check-in facilities are provided at Hong Kong and Kowloon stations. They were constructed and are operated by the MTR Corporation Ltd. (MTRCL). Much of the capital funding for MTRCL comes from the sale of development rights over or adjoining many of the stations. The design and construction of a station thus creates a development platform as well as a public transport facility. Where development is to be integrated, the station structure has to be designed to carry the imposed loads, and the masterplans for station areas have to provide access and servicing to the buildings above as well as the station.

The MTRCL capital expenditure programme has always been partially funded through joint venture development over and around stations. Prior to construction of the rail link to the new airport, stations built on the system over the previous two decades supported 31,366 apartments, 193,600 square metres (2 million square feet) of offices, 310,400 square metres (3¼ million square feet) of commercial space and 139,000 square metres (1½ million square feet) for government institution and community use.

Apart from the Ground Transportation Centre at the new airport there are three new stations on the AEL. They represent three very different types of urban metro station and illustrate many of the issues involved in urban rail construction.

Hong Kong station is the terminus of the AEL and is constructed on reclaimed land alongside the heart of the central business district on Hong Kong Island. Links to MTRCL's existing Central station on the Island and the Tsuen Wan lines were necessary for passenger interchange. This required construction of subways and passages underneath the Pedder Street underpass and below the existing high-rise buildings of Exchange Square.

Previous page: The subway links the new Hong Kong station to the Central Station on the Mass Transport Railway.

Opposite: Entrance concourse to Hong Kong station.

The station at Kowloon is built on a major reclamation area on the western side of the peninsula. This is effectively a green-field site for there were no existing buildings affecting the route of the railway or the site of the station. As well as the new station, the reclamation accommodates the northern portal and toll plaza of the Western Harbour road tunnel and major road infrastructure.

Whereas Hong Kong and Kowloon stations are underground, Tsing Yi station is elevated, within a seven-level commercial podium supporting 12 high-rise residential towers, on the island of Tsing Yi.

In addition to these AEL stations, there are new stations on the local railway, the Tung Chung Line (TCL). Tung Chung station is the western terminus for the service and is located in the development area on Lantau Island south of the airport reclamation. The new TCL stations are planned with developments around them. This integrated planning approach capitalizes on the commercial potential of these sites due to high passenger flows through the stations.

The two separate railway lines provide services for different functions. The AEL provides a fast link from Hong Kong, Kowloon and Tsing Yi to the Ground Transportation Centre at the airport, with check-in facilities for airport passengers available at Hong Kong and Kowloon. Trains travel at up to 135 kilometres (85 miles) per hour, and journey time from Hong Kong to the airport is 23 minutes. The local TCL service serves two additional stations – Olympic and Lai King, between Kowloon and Tsing Yi – as well as the three AEL stations and the terminus at Tung Chung. Lai King station has an interchange with the MTRCL's Tsuen Wan line. At the stations serving both AEL and TCL, the lines are served by separate platforms, with those of the AEL above the TCL platforms, but tracks are shared for much of the route.

Hong Kong's MTR is a fairly recent network, with the first line opened in 1979. It therefore has many more modern features than the London Underground system or the New York subway. Stations are air-conditioned and well planned within the constraints. The trains do not have divisions between the carriages, so passengers can see and walk through the train easily (crowding permitting!). As well as increasing the capacity of the train, this provides a more secure environment, and comes as a pleasant surprise for passengers more used to metro systems built over a century ago.

A system-wide set of generic design principles was established by the MTRCL to give a degree of uniformity to the stations. These principles included glass-enclosed lifts to help maintain an open aspect to the stations and platform screens with doors to assist in the control of environmental conditions on the platforms and concourses. The platform-screen doors required careful design to ensure they were sufficiently robust but light enough to operate smoothly and reliably and to minimize the kinetic energy released when they close, thus reducing the risk of injury. They were glazed to permit viewing of the approaching trains and the advertising panels on the tunnel wall, and lighting levels were set to prevent reflections obscuring the advertising. The characteristics of the glazing and the ventilation system had to be considered to prevent condensation: the tunnel air is humid and can be as hot as 40°C (104°F), whereas the platforms are air-conditioned to about 24°C (75°F).

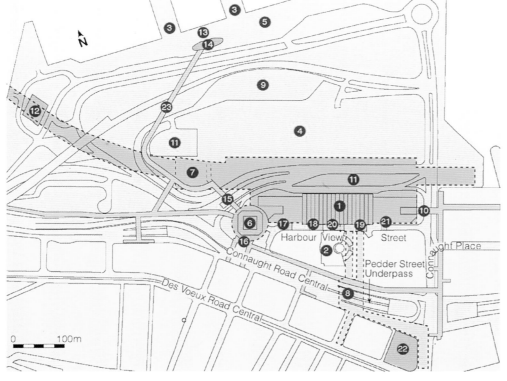

Key

1	Hong Kong station
2	Exchange Square
3	Ferry piers
4	Northern development site
5	Promenade
6	International Finance Centre
7	Underground chiller area
8	Subway
9	Bus station
10	New footbridge J
11	Temporary car park
12	Hong Kong ventilation building
13	Cooling water pumping station
14	Hong Kong power supply building
15	New footbridge C
16	New footbridge N
17	New footbridge O
18	New footbridge P
19	New footbridge O
20	New footbridge X
21	New footbridge R
22	World Wide House and below ground concourse for Central Station
23	Temporary footbridge
	Indicates extent of Station basement and tunnels

Opposite: Hong Kong station was constructed on reclamation in front of the 50-storey Exchange Square complex, formerly on the waterfront. Subways and footbridges create links to Central Station on the Mass Transit Railway and into and through buildings fronting Harbour View Street.

Above: The ticketing area inside the atrium also provides for airline check-in, with checked baggage being carried in a separate compartment of the Airport Express trains.

Hong Kong station

The station, constructed in the new reclamation on the north side of Hong Kong Island, forms the southern terminus of both the AEL and the TCL. The reclamation created 20 hectares (50 acres) of land and extends up to 350 metres (1,150 feet) from the existing shoreline. It not only provides land for the station but also allows for extension of the Central business district. Six new ferry piers and a new bus terminus enable interchange with the railway lines and bus/ferry interchange, as well as access to the area. The station occupies 4 hectares (10 acres) and is the arrival point for passengers coming from the airport into Hong Kong itself, particularly businessmen coming into the city and the business district.

The Hong Kong station had to relate to existing development and optimize the site value generated through property development on the site. The reclamation and new station were constructed on the harbour side of Exchange Square, whose 53-storey office towers were marketed as 'the last waterfront in Central' when they were built. Harbour View Street, which was originally along the waterfront, forms the southern boundary of the station site but no longer has a view of

the harbour. The station is, therefore, at the heart of a new development district, and its pedestrian and vehicle accesses had to recognize the needs of any future expansion of the area. As well as being responsible for the detailed architectural design of the station itself, Arup Associates and Rocco Design Partners were appointed to masterplan the whole reclamation site. Given that the station is also the in-town airport terminal for Hong Kong Island, there had to be access and kerb space for public and courtesy buses, limousines and taxis so that they could pick up and set down close to the check-in hall. The conflicting demands of movement and development had to be reconciled and service and access requirements fully planned throughout the area.

Design concept

Because the station is at the heart of Hong Kong's business district, there was a desire on the part of the MTRCL to create a statement that reflected the importance and prestige of the location. The MTRCL's objectives for the station – and, in fact, for all the new stations – included:

– an attractive overall design
– exploitation of natural lighting
– clarity of orientation and circulation
– durable and easily maintained finishes

With the AEL platforms located above those for the TCL, a key aim of the design was to produce open and attractive spaces. There are underground concourses at three levels: the upper and lower station-platform levels and an intermediate level with access to a subway under Exchange Square direct to Central station. Phase 1, completed to coincide with the opening of the airport, provided a single platform for the AEL, for both arrivals and departures. A second platform will be commissioned as part of the Phase 2 development in 2003. The TCL has stair, lift and escalator access down to an island platform with two platform faces.

Pedestrian links

At first-floor level a series of five footbridges join Harbour View Street and Exchange Square, providing continuity for the pedestrian routes through the main circulation level of that development. A further footbridge at the eastern end of the station connects with Hong Kong's main Post Office and the elevated walkway linking the ferry piers with the main business area in Central district. At the western end, pedestrians using the elevated walkway along Connaught Road Central now pass through the public lobby area of the new International Finance Centre 1 to the station development site.

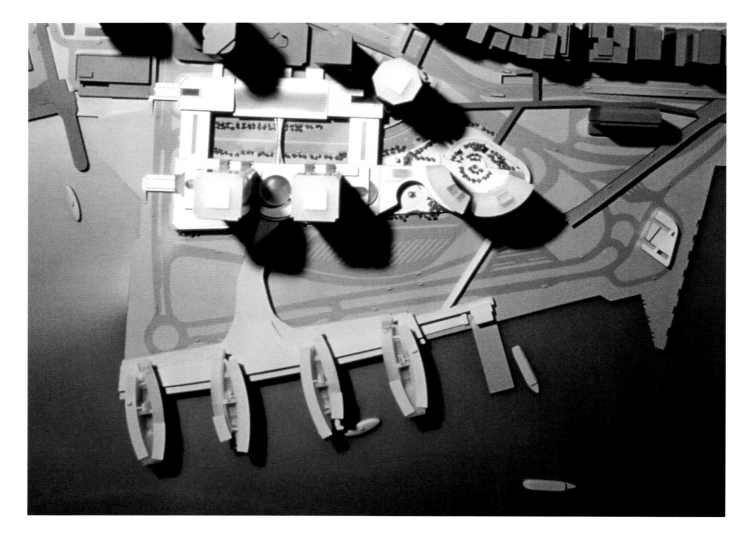

Opposite: The reclamation provides an integrated transport terminal with a bus station and new ferry berths. The space over the station is being developed for high-rise commercial development to help fund the railway.

Above: The steel roof has a wave form, a popular style in modern transport terminals.

Architectural features

The main feature of the check-in hall is the glazed north wall of the atrium. Check-in is at ground level with four levels of retail development at the rear and either end of the concourse, opening out onto the atrium. The roof has a wave form, and finishes are light in colour with cool grey and white surfaces and occasional splashes of MTR dark blue. With these details the atmosphere of the main entrance is more like that of an airport terminal building than a mass transit railway station. This is not surprising for the check-in area fully replicates the check-in facilities provided in the departures area of an airport. The colour scheme is continued below ground in the concourses and on the platforms, where it contrasts with the expanses of mosaic tiles and dark ceilings used on the older MTR lines.

Light wells in the check-in hall provide natural light for the AEL departure platform below. The TCL station underneath this has a concourse at the intermediate level, from which a subway under Exchange Square allows passengers to use the travelator or walk to Central station on the MTR Island and Tsuen Wan lines.

Construction

The station was constructed by the top-down method within a box formed by a perimeter diaphragm wall. An advance contract with the reclamation contractor was agreed for part of the diaphragm walling and bored piles; otherwise there would have been delays to the programme. The timing was critical, for with the railway forming a main access mode for air travellers it was essential that the station be open at the same time as the airport.

The steel roof members of the main hall are constructed from standard sections and hidden by a false ceiling which has access ways for maintenance above it. The roof slab of the underground station box provides support for the external roads and lay-by areas, as well as being the floor of the main hall. Two metres (6½ feet) of soil were placed over the roof slab in the traffic circulation areas so that buried public utility services could be accessed from the road level without disturbing the MTRCL-owned railway structure.

Originally, the subway under Exchange Square and the Pedder Street underpass was to have been constructed as a bored tunnel, but ground conditions and the congested site resulted in the use of cut-and-cover methods. Constraints included old seawalls, existing buildings, the Pedder Street underpass and utilities plant. Harbour View Street had to be diverted to construct the subway and an advance works contract was let because the diversion could not take place after the station excavation had started. This section of the subway was between diaphragm walls 11 metres (36 feet) apart, and a 1.3-metre (4-foot) slab was required for road reinstatement. The concourse of the existing Central MTR station is underneath World Wide House at the junction of Pedder Street and Connaught Road. The importance of this junction meant it was not possible to close it completely to road traffic, so a complex traffic-management plan involving lane closures was implemented.

The subway works contract included a chiller building, a ventilation building, seawater-pipe tunnels, a power building, retail development structures, footbridges and roads as well as the cut-and-cover tunnels.

Opposite: The Tung Chung line concourse, which does not have the light wells that benefit the Airport Express Railway concourse in the upper basement level, has illuminated ceilings to create an attractive environment.

Below: Construction of the subways beneath the buildings around the Hong Kong station site involved a variety of construction techniques.

Below: The station is in the heart of a new development zone.

Opposite: The new road network provides access to the station and development at several levels.

Kowloon station

Kowloon, the largest of the AEL stations, is the centrepiece of a major reclamation on the western side of the Kowloon peninsula. The site is remote from other development and so the problems of working under and around existing roads and buildings – as experienced on the Hong Kong station site – were not encountered. The station and related development occupies 13.6 hectares (34 acres), a small part of the total reclamation, which extends to 334 hectares (835 acres) and will include housing for over 150,000 people.

The station was nevertheless a complex planning task for it was integrated with the masterplanning of the development, including 22 high-rise buildings over and around the station. It is predicted that once the developments are completed, 8,300 AEL passengers and 44,000 TCL passengers will use the station during peak periods.

Access

Good road access is important, because the station provides check-in facilities for passengers from the whole of Kowloon. Facilities are therefore provided for public and hotel buses, limousines and taxis. The internal road network surrounds the platforms and concourse areas and can accommodate 12-metre (40-foot) coaches. As well as provision for private coaches and limousines there are two drop-off kerbs for private use, each some 200 metres (660 feet) long, and taxi pick-up bays to cater for 500 taxis an hour. Taxi flows of this size need special control systems to direct both passengers and vehicles to vacant bays, since the demand is much higher than can be accommodated by a traditional taxi rank.

Arup Transportation have undertaken research and field trials in the UK and Hong Kong to develop taxi-rank designs for a number of projects where high volume was predicted, including Paddington and King's Cross/St Pancras railway stations and Heathrow Airport in London, the Ground Transportation Centre at Hong Kong Airport, and Hong Kong station and this one on the AEL. For Kowloon, a taxi station was devised with air-conditioned queuing areas and room for simultaneous loading of five taxis on each side of two piers.

Planning

The Kowloon station project included cut-and-cover tunnels either side of the station and the Kowloon Ventilation Building at the northern end of the MTRCL's immersed tube tunnel under the harbour. The station and its future development area form a major piece of urban architecture. Terry Farrell & Partners were appointed for architecture for these projects and for masterplanning of the whole 13.6-hectare (34-acre) site, with residential, commercial, retail and hotel developments. Their proposal was significantly different from those of other bidders. The form and geometry of the concept study were completely reorganized and simplified so that all station functions could be contained within a box 300 metres by 180 metres (984 feet by 590 feet). This gave order, symmetry and repetitive modularization to the function and structure. The overall external form of the station box itself was not an overriding issue since most of the structure is below ground; any that is visible provides a platform for future development.

The planning logic can be seen from the clear positioning of the major station functions:

- TCL platform at the lowest level
- AEL platform above the TCL
- departures level, with large internal drop-off road and in-town check-in facilities for passengers and baggage
- arrivals level with associated transport functions: taxis, buses, minibuses, private car parking and internal departure pick-up road
- station entry from the developments
- extensive service zones in each of the levels, as required for their function

Tower 1

Tower 20

Tower 21

Tower 19

Tower 18

Tower 3

Tower 4

Tower 5

Tower 6

Tower 7

Tower 8

Tower 22

Tower 23

Tower 17

Tower 16

Tower 15

Tower 9

Tower 13

Tower 10

Tower 14

Tower 12

Tower 11

Opposite: Within the development area, access points to the station for pedestrians are set in landscaped parkland.

Above: The masterplan for the development provides for 22 high-rise blocks to create a new town above the station concourse.

Above: A complex group of cut-and-cover concrete boxes carries four lines (one in each direction for the two railways), refuge sidings for crippled trains, ventilation plant and services. The cross-section changes where the two railways share tracks in the tunnel under the harbour.

Opposite: The escalator groups link the upper concourse with separate platform-concourse levels for the Airport Express Railway and the Tung Chung line. The well provides natural light to both lower-concourse levels.

Construction method

Arup proposed an open-cut method of excavation, thus avoiding the need for costly diaphragm walling, and this was the approach adopted by MTRCL. This method had several advantages:

– Construction was not reliant on a small number of diaphragm-wall contractors with commitments on many other airport projects during the same construction period
– Durability is a prime consideration in this aggressive environment, and in-situ construction in open cut gave greater quality control
– Element planning is freed up, allowing structural freedom to use perimeter walls as deep beams and load-distribution elements

To enable construction the water table had to be lowered by well-point dewatering to a level of -18 metres (-59 feet). Because of the high water table, rising to the surface under extreme conditions, the station box had to be tied down to counteract large uplift loads on the basement structure. The piles therefore had to be designed for tension as well as compression.

There are three suspended levels above ground accommodating retail and commercial facilities for passengers en route to the station. The check-in and baggage-handling facilities are below ground between the platforms at both ground and basement levels in the central part of the platforms, with plant rooms at either end.

Foundations for the proposed developments over the station were incorporated in the station structure. The bored piles on which the station is founded vary in length from 20 to 106 metres (66 to 350 feet).

Architecture

The station is a major building for it accommodates the facilities of an airport departure area and arrivals concourse as well as providing access to the railway platforms. The station and the development form a core for the future development of a whole new district.

The idea for a central square above the station was based on the traditional concept of public space as the defining element in urban design, updated to suit the dense conditions of Hong Kong.

The square, a podium-level space between buildings, is also the roof to the station and the location of the entrance into the great halls of the airport railway. The entrance is in the exact centre of the station, at the crossing-point of the rail and station-concourse axes, and acts as a gateway to both. On the rail axis is the entrance to the new city to be built above the station. On the concourse axis there are two entrances: to the AEL Departures Hall to the west and to the Arrivals

Hall and TCL to the east. Placing the main entrance in the centre of the square ensures that the large columns needed to support the development above do not intrude on the busiest spaces in the station concourse.

The concourses are designed as a series of interlinked halls within the rectilinear grid of the station box, with each hall containing one of the principal functions of the station. The folded profiles and angled planes of the ceilings create a separate identity for each space.

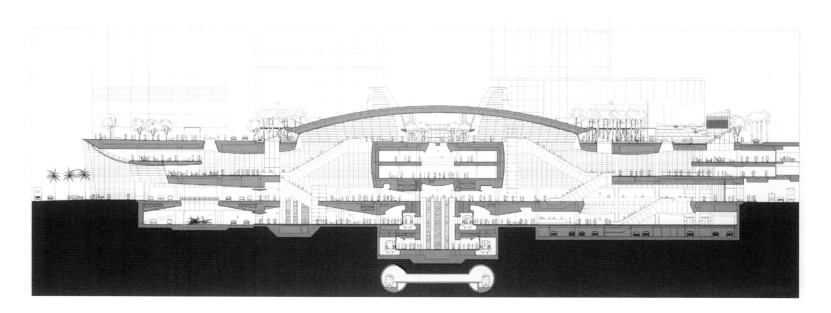

Opposite: The ceiling helps to define the functions of the various parts of the main station concourse.

Above: The cross-section through the station box reveals the various concourse levels.

Opposite: The Ventilation Building provides air to the station and the tunnel under the harbour, and has a sea-water intake for cooling.

Below: The air intakes of the building create a landmark on the southern tip of the West Kowloon reclamation, visible at the entry to the harbour.

The Ventilation Building

The ventilation plant at the northern end of MTRCL's immersed tube is contained in a landmark structure in the planned new park. It is highly visible from viewing points on Hong Kong Island and from vessels approaching Victoria Harbour, which lies between Hong Kong Island and Kowloon. The building provides plant to dissipate the heat generated by trains, passengers and plant, using seawater cooling. It also provides air to ventilate the running tunnels and provides pressure relief from the piston effect of trains running through them.

The organic form of the building is intended to provide a sculpted landscape to enhance the park, and derives from a careful analysis of the functional profile of its operations. A single curved form envelopes all parts of the building, a low-cost reinforced-concrete structure finished in simple materials. The bright colours and animal-like form have led to a variety of pet names for it, including 'the whale', 'the wave', 'the dragon', 'the grasshopper' and 'the sail boat'.

Tsing Yi station

Tsing Yi is the only one of the major stations located above ground level. It lies on the eastern side of Tsing Yi Island, on a site reclaimed in the 1980s and used as a lorry park prior to construction of the railway. The station is near the midway point between the airport and Hong Kong station.

The railway crosses the Rambler Channel from Kowloon on a bridge, and continues over viaducts to the west of the station into the Tam Kon Sham tunnels in the centre of the island before emerging onto the Tsing Ma bridge to the west. The Tsing Ma bridge is the longest suspension bridge in the world to carry both road and rail traffic, with a main span of 1,377 metres (4,518 feet) and a total length of 2.2 kilometres (1½ miles). The bridge carries two railway tracks at the lower level and six highway lanes at the upper level. The Rambler Channel rail bridge carries two tracks for the AEL and two TCL tracks, and the station has the same configuration. West of the station is a length of multi-track, multi-level viaduct structure which carries the viaduct structure above the at-grade road network and the Tam Kon Sham interchange.

Apart from the seawater pump house, the station is constructed entirely above ground. The podium has seven floors (including ground level), with the tracks from Hong Kong directly above the tracks to Hong Kong. There are residential towers above the podium, which rises 34 metres (112 feet) above ground level. The platforms are at the second and third levels, with 1,350 parking spaces on the fourth and fifth decks. The building accommodates 60,000 square metres (646,000 square feet) of commercial and retail space at the platform levels and the MTRCL Operations Control Centre, a triple-height 6,000-square-metre (64,600-square-foot) facility from which both lines are controlled.

The residential and commercial development comprises 291,000 square metres (3 million square feet) of gross floorspace. The station podium is the platform for 12 residential blocks which provide 3,500 apartments for a population of some 10,000.

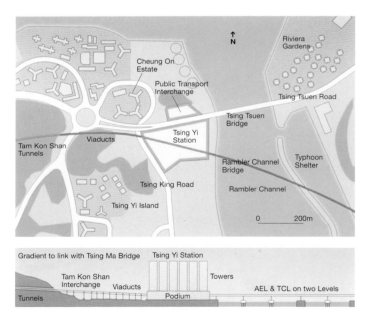

Left and opposite: The station is between tunnels through the centre of Tsing Yi Island and the Rambler Channel bridge to the south. A separate track provides access to the rolling stock depot for both the Airport Express and Tung Chung lines.

Design

The entrance to the station is on the north side and has a five-storey atrium with extensive glazing to the north and north-east elevations. There is a curved roof above the AEL departures concourse area. The north-east elevation provides views across the Rambler Channel. The internal finishes are white, grey and powder blue, similar to the other stations on the railway. The structure is in reinforced concrete with round columns. Those supporting the podium are 1.2 metres (4 feet) in diameter and those supporting the towers are 2.4 metres (8 feet) in diameter. The towers are grouped on each side of the railway, to avoid the need for large supports passing through the station area.

Pedestrian access includes three footbridges, two connecting with the transport interchange, a 9,500-square-metre (102,000-square-foot) single-storey reinforced-concrete building on a separate site to the north of the station. It includes an open bus terminus on the roof, a lorry park below at ground level, and an adjoining area for minibuses and taxis. There is vehicle access direct from the existing Tsing Tsuen road bridge, from which a slip road at fourth-floor level provides access to residential parking on the fifth level and the AEL departure-concourse drop-off area on the third floor. Access to the top of the podium is provided by spiral ramps for delivery trucks and emergency-service vehicles. The passenger flow into and out of the station is initially estimated to be 17,000 per hour, rising to 24,000 in 2021.

Below: The site for the station was an undeveloped area of reclamation that had previously been used as a truck park.

Opposite: Tsing Yi, the only station completely above ground on the airport railway, provided the opportunity for a spacious concourse area. From here there is access to the commercial development serving the new housing blocks above the station.

The commercial centre comprises four levels, from the ground to the third floor in the north-eastern area of the podium. It has a separate identity but shares access with the station complex. The main floor is at first-floor level, providing access via footbridges and walkways from surrounding areas. There is a landscaped promenade alongside the Rambler Channel linking into the ground floor. The design allows natural light to penetrate into all areas of the centre which, with 46,000 square metres (495,000 square feet) of food outlets and entertainment facilities as well as shops, aims to become a major attraction in Tsing Yi town centre.

The approach viaducts

The viaducts at the western end of the station are 420 metres (1,380 feet) long and up to 25 metres (82 feet) above ground level. They have varying alignments, cross-sections and profiles, for the tracks merge and diverge in this area. The original Arup submission considered containing some of the tracks in box-girder structures, an approach which had been developed for the Bangkok Elevated Road and Train System, but an alternative design using open viaducts with acoustic-screening concrete parapets where required was finally adopted. Close to the residential towers, steel portal frames carry transparent acoustic panels over the tracks to screen the apartments from noise.

The viaducts are carried on portal frame structures which can accommodate the variations in width of structure and span over tracks and the various ground-level constraints, including major roads.

Tung Chung station

At the Lantau Island end of the line there are separate termini for the AEL and the TCL. The AEL terminates in the airport's Ground Transportation Centre, which adjoins the main terminal building on the artificial island constructed for the airport. Tung Chung station is on Lantau Island, south of the sea channel that separates the airport from Lantau. The railway approaches Tung Chung from the east at ground level, with the AEL and TCL sharing two tracks. The lines separate east of Tung Chung with the AEL looping round alongside the North Lantau Highway to cross the sea channel to the airport and the TCL dropping into a cut-and-cover tunnel to the underground terminus. The line is designed for a future extension to Tung Chung West and possibly a loop continuing to the north underneath the airport to provide another connection to the mainland.

Opposite: The concourse has a large wing-like roof, reflecting the concept of flight and providing the maximum amount of light.

Below: The masterplan for the new town of Tung Chung has the station at its heart.

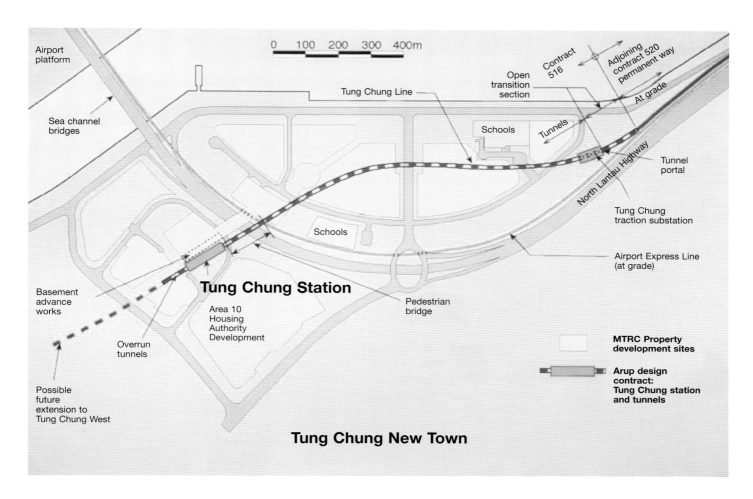

0 100 200 300 400m

Airport platform

Sea channel bridges

Tung Chung Line

Open transition section

Contract 516

Adjoining contract 520 permanent way

At grade

Schools

Tunnels

North Lantau Highway

Tunnel portal

Tung Chung traction substation

Schools

Tung Chung Station

Pedestrian bridge

Airport Express Line (at grade)

Basement advance works

Area 10 Housing Authority Development

MTRC Property development sites

Overrun tunnels

Arup design contract: Tung Chung station and tunnels

Possible future extension to Tung Chung West

Tung Chung New Town

Town-centre planning

The station is integral to the Tung Chung development which was planned as an industrial/housing support community for the airport. The overall site comprises 760 hectares (1,900 acres) of natural and reclaimed land and is expected to have an ultimate population of 320,000. The site for Tung Chung station and associated development is 21.6 hectares (54 acres). The development comprises 843,000 square metres (9 million square feet) of which 750,000 square metres (8 million square feet) is residential.

Station design

Although the station is the terminus of the line, the platforms are designed as through-platforms to allow for future extension of the line. There is a central island platform with access by lift, stairs and escalators from ground level. Plantrooms are above the running tunnels, and station staff, concession and ticket areas are above these at ground level. Because the site is close, both visually and functionally, to the airport, the design aimed to reflect the concept of flight and incorporate as much light as possible. The concourse is covered by a large wing-like roof, cantilevered from a double row of tall, tapering columns. A clerestory roof provides natural light in the passenger areas of the station. Across the concourse and station forecourt is a high-level walkway linking to the commercial development and nearby housing.

Above: The lifts and other details for the whole of the new railway were detailed to a common design and colour framework.

Opposite: As with all the new stations, convenient interchange to other forms of transport is provided as part of the integrated transport strategy.

Conclusion

The Hong Kong airport railway provided a wide variety of design challenges. The solution to the simple requirement of a high-speed link to the airport was one of enormous complexity in that it had to meet the needs of both local and airport traffic. In-town check-in was an integral part of planning the airport, reducing the demand for car and taxi traffic at the airport itself. However, this meant that each of the main stations had to take on aspects of an airport terminal, meeting high-specification demands of customer comfort and passenger and vehicle access. They therefore required the major mechanical and electrical services expected of an airport, as well as those of a passenger railway. In Hong Kong, with high land values and the requirement to provide access and support for development platforms, provision of these services poses a particular problem for engineers and architects. The completion on time of a railway of such complexity and with such high standards of design is a credit to the entire team of architects, engineers and contractors, and owes much to the strong project-led management team put together by the Mass Transit Railway Authority.

> Channel Tunnel Rail Link

South-east England, UK, 1989–2007

The Channel Tunnel between Britain and France was opened in 1994. It is used by shuttle trains carrying cars and trucks between terminals at each end, long-distance freight trains, and Eurostar passenger services. Most Eurostar trains run between London, Paris and Brussels, with some extra trains from London to EuroDisney, Avignon and Geneva. The trains are specially designed to cope with the three different electrification systems in France, Belgium and England, and are capable of travelling at 300 kph (186 mph).

French Railways have had a programme of building new lines for high-speed trains for some years, and completed one between Paris and the Tunnel via Lille, prior to the Tunnel opening. Belgian Railways completed a high-speed line from Lille to Brussels a few years ago. In France and Belgium, therefore, the trains were able to operate at their maximum speed. In England, however, they were required to use pre-existing lines between London and the Tunnel, and the track, signalling and power supply had to be upgraded to allow them to run at a maximum of 160 kph (100 mph). A maintenance depot was built at North Pole, in west London, and a new station, Waterloo International, was built alongside the existing mainline station at Waterloo with facilities to accommodate the long Eurostar trains.

A study in 1987 by British Rail (BR) concluded that a new line was needed to provide the capacity and speed necessary to exploit the opportunities created by the Channel Tunnel. BR chose a route that followed existing roads and railway lines through the rural sections of Kent, approaching London from the south east. Companies were invited to bid to be BR's partner in developing and constructing the route and Eurorail, a consortium of Balfour Beatty and Trafalgar House, was chosen in 1989. BR set up a joint team with Eurorail to manage the project.

Arup had submitted a proposal, which did not conform to the terms of reference, based on an alternative route approaching London from the east. This was designed to reduce environmental impact by following existing roads and railway lines, and provide intermediate stations in areas requiring regeneration. The line would be used by express domestic services as well as European trains, and terminate at Kings Cross/St Pancras. Eurostar trains would also be able to continue using Waterloo International via a spur in the Dartford area linking to existing BR tracks. The Thames would be crossed in a tunnel near the existing Dartford road crossing; the crossing and its approaches would have four tracks to enable freight and local trains to cross the Thames east of London, in addition to the high-speed trains.

Opposite: A Eurostar train, capable of 300 kph (186 mph), passes the viaduct being constructed to separate through trains and those stopping at Ashford International Station.

The Arup route was seen as promoting the regeneration objectives of east London as well as minimising environmental impact and attracted wide support, particularly amongst local authorities in the areas that would benefit.

Arup also proposed excluding contracting companies from the client partnership until they were appointed for specific work contracts. This was because Arup considered that the experience of the Channel Tunnel, where the contractors led the bid and created a client company that they controlled, did not provide a sufficient discipline to optimise the design and control costs and programme.

The Arup bid was rejected in 1989, but the company remained convinced of the benefits of its proposal and carried on developing the concepts at its own expense. Arup lobbied for its route, the government had second thoughts, and in October 1991 announced that it would abandon the route BR had been working on and develop another along the lines of Arup's proposal. Arup joined the project team, which became an agency company of BR, called Union Railways Ltd (URL). Following consultation and research it was decided to construct a two-track high-speed passenger railway, which would have the capacity to accept European intermodal freight vehicles if required. The additional train weight of the freight trains required the controlling gradient to be $2^{1}/_{2}$ per cent, compared

with $3^{1}/_{2}$ per cent on the French high-speed network, which is exclusively for passenger services.

Between 1992 and 1994 URL refined the eastern corridor into an alignment that would form the basis of a Bill in the UK Parliament giving the necessary powers to construct and operate the line when enacted. A station was envisaged at Ebbsfleet, on the south bank of the Thames near Dartford, as proposed in the Arup scheme. It was left to the concessionaire to decide whether to include the option of a station at Stratford, in east London on the north side of the Thames.

At the same time the government reconsidered the financial and ownership structure to be used to deliver the line. It concluded that a Design Build Finance and Operate (DBFO) concession would be let, with the private sector bid being assessed on the willingness of the bidder to accept risk and the amount of government contribution required. Passenger train operations through the Tunnel were to be operated by a joint company, owned by the British, French and Belgian state railways. It was decided that the existing part of the passenger train operation owned by British Rail – European Passenger Services Ltd (EPSL) – would be transferred to the successful bidder.

Arup was instrumental in assembling a consortium to bid for the concession; London and Continental Railways (LCR) contained engineering design and project management, transport operations and financial skills. The designers and project managers were Arup, Halcrow, another UK-based international consulting engineer, Bechtel, the American project managers, and Systra, a consultancy providing the railway engineering and systems expertise of the French national railways (SNCF) and the Paris public transport operator (RATP). These companies formed Rail Link Engineering (RLE) as a joint venture. The operators in the group were the airline owner Virgin, and the airport, bus and coach operator National Express. S. G. Warburg provided the banking skills, and the consortium also included the utility London Electricity.

Bids were invited in February 1994 and four of the nine applicants were prequalified that July. Full proposals were submitted in March 1995, and LCR and Eurorail were shortlisted to enter the final stage of the competition. Final offers were submitted in December 1995 and LCR were selected. Transfer of the project, including URL and EPSL, was effected in May 1996. LCR included the Stratford station in their bid and that became part of the project. The Bill received Royal Assent and became law in December 1996.

LCR proceeded with fundraising, but the market was concerned about risk, particularly as the Channel Tunnel had been late and over budget, and so LCR was forced to approach the government for support in January 1998. In 1994–96 British Rail had been split up and sold to the private sector, the track and signalling being transferred to Railtrack, a newly established company that was the subject of a public flotation. A financial restructuring was agreed, with Railtrack agreeing to purchase Section 1 of the Channel Tunnel Rail Link (CTRL) on completion, with an option on Section 2. The government agreed to back £3.75 billion of the private debt. Section 1 runs from the Channel Tunnel to Fawkham Junction, east of Swanley in Kent, with Section 2 completing the route to St Pancras.

With the funding in place, work on Section 1 was started in October 1998, for completion in 2003. Section 2 construction commenced in July 2001 and is due to be completed in 2007. When Section 1 is complete there will be a link via a previously abandoned line with the existing rail network near Swanley, from where Eurostars will travel over existing tracks to Waterloo.

Opposite: The Medway Viaduct, at 1.2 kilometres ($^3/_4$ mile) long and with a central span of 152 metres (500 feet), is the longest high-speed railway crossing in Europe.

Below: The CTRL rises from the Thames Tunnel and passes under the approach viaduct for the Dartford Bridge, which carries M25 traffic over the Thames. The cable-stayed bridge was completed in 1992.

Benefits of the CTRL

Journey times from London to European destinations will be reduced, high-speed domestic services will be able to operate from East Kent to London, existing capacity on suburban lines will be increased by removal of the Eurostars and more freight paths will be available to the Channel Tunnel. In addition it is expected that the new stations at Ebbsfleet, Stratford and St Pancras will act as catalysts to regeneration in these areas.

Eurostar's current 7 million passengers per annum (mppa) will benefit from the reduced journey times. The improvements can be expected to generate more passengers and increase the competitiveness of the railways with air services. Section 1 will reduce journey times from Waterloo to Paris from 175 minutes to 155 minutes, and to Brussels from 160 minutes to 140 minutes. When Section 2 is complete, journey times will be 135 minutes from St Pancras to Paris and 120 minutes from St Pancras to Brussels.

Procurement of the CTRL

LCR appointed RLE as designers and project managers for the whole of the works. RLE operates as an integrated project team in its own offices with staff seconded from the four partners. The works were divided into four geographical areas, with each of these subdivided into several contracts. A fifth series of contracts – comprising electrification, signalling and permanent way works – was let for the whole project.

RLE has a very wide-ranging responsibility, with an EPC (engineer-procure-construct) remit to manage and deliver the total project on behalf of LCR. On such a complex project, managing the design is difficult enough: many disciplines have to be pulled together. These include architecture, building structures and services, environmental assessment, highways, utilities, foundations and earthworks, bridges and structures, permanent way, overhead line electrics, signalling and communications.

The procurement strategy was to choose contractors before the design was complete, involving them as it was finalized. RLE elected to use the New Engineering Contract as the basis for contracts, generally using the Target Price form with the levels of 'pain/gain' incentivization being set to maintain margins relative to the percentage fee of the various market sectors (civils, building, rail). For smaller contracts where the design was complete and the risk of change small or entirely covered by a third party, the Lump Sum form has been used to achieve competitive prices and minimize the cost of contract administration, for both the contractor and RLE, to the client's advantage.

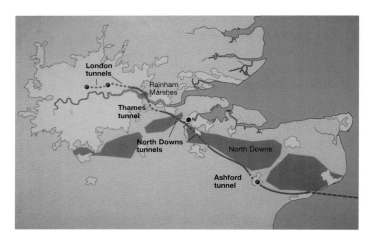

Above: The route of the new railway line was designed to avoid zones of high quality landscape and serve regeneration areas.

Above: A tunnel reduces the impact of the railway on the North Downs area of Outstanding Natural Beauty.

Opposite: Stratford International station, constructed in a box, will provide interchange with underground and surface lines, and the Docklands Light Railway.

Stations

Two new intermediate stations are to be constructed on the CTRL, at Ebbsfleet and Stratford. In addition, major alterations were necessary at St Pancras and Ashford. The historic station at St Pancras required extensions and changes to accommodate both the Eurostar trains and those of the Kent Express line, for which it is the London terminus. This work is discussed elsewhere in the book (see pages 126–133).

Ashford

CTRL trains will use the existing station at Ashford International, which provides access to international trains for passengers from the Kent area. Work has been carried out in the area to allow those not stopping at Ashford to bypass the station. Ashford will also be the point at which Kent Express trains from Margate and Canterbury join the CTRL, reducing journey times to London by up to an hour.

Stratford

Stratford is the existing major transport interchange in east London, and in future will also provide many connections for international passengers. One of its key links is to the Canary Wharf area, a major financial centre constructed over the last ten years in London's Docklands.

Currently serving Stratford are two Underground lines (the Jubilee and Central), the Docklands Light Railway, the main lines from Liverpool Street station to East Anglia (connecting with ferry services to Holland and Scandinavia from Harwich), and the North London Line (NLL). The latter, an orbital service avoiding central London, terminates at Richmond in west London. It also serves London City Airport to the south-east of Stratford.

When works are completed to increase the capacity of the Lea Valley line it is expected that direct services will run to Stansted Airport, London's third and fastest growing airport, now serving 15 mppa.

A new cross-London link from Liverpool Street to Paddington, providing through services from Stratford to Heathrow Airport and Reading to the west, will further increase the accessibility of Stratford. Outline design for this project is under way for completion in 2012–15.

Together with St Pancras (on the north side of central London) and the existing terminus at Waterloo (which serves local trains from the south and west), the new station at Stratford means that easy access to international services is available from a wide range of transport links.

A large area of former railway land at Stratford, granted to LCR as part of the concession, provides an opportunity for major commercial and residential development close to the international station. The area may also be the focus of a London bid to host the 2012 Olympic Games.

Stratford station is being constructed underground in a cut-and-cover box 1.1 kilometres (³/₄ mile) long. Both international and domestic platforms will be provided. A surface connection will allow future depot facilities to be built in the area (in addition to those at North Pole in west London) and allow Eurostar trains to be routed via the NLL to the East Coast Main Line (ECML).

The Stratford station box is the starting point for the London Tunnel drives towards St Pancras. The London Tunnel runs underneath the NLL, avoiding disturbance to property above. The spoil from the tunnel will be used as fill on the development areas, avoiding the need to remove it from the site.

Ebbsfleet

The proposed Ebbsfleet station is south of the Thames on the fringe of Greater London. This station will be above ground in an area of extensive chalk workings for cement manufacture. It will serve the largest urban regeneration area adjacent to London, as part of a plan to strengthen the economy of the Thames Gateway, a relatively depressed region compared with the Heathrow area to the west of London. Ebbsfleet will be easily accessible from the M25, London's orbital motorway, and will have around 9,000 parking spaces. It will also be the point at which Kent Express services from Rochester and Chatham join the CTRL. A station for the North Kent Line will be built adjacent to Ebbsfleet International.

Below: Ebbsfleet station (both pictures) will provide an interchange between traffic on the M25 London Orbital Motorway and the Eurostar and Kent Express trains.

Opposite: The tunnel shield is being installed in the Stratford station box for the first London Tunnel drive towards St Pancras.

Route civil engineering

The total route length from the Channel Tunnel to St Pancras is 109 kilometres (68 miles). Section 1 is 74 kilometres (46 miles) and is estimated to cost £1.9 billion. Section 2 is 39 kilometres (24 miles) and is estimated at £3.3 billion. The higher costs for Section 2 reflect the fact that a greater proportion of the route is in tunnel as the line approaches London, and the building of the stations. The works are estimated to create 8,000 construction jobs. There will be 152 new bridges – 60 carrying railway lines, 62 carrying roads and 30 carrying footpaths. Total excavation is likely to approach 20 million cubic metres (26 million cubic yards).

Section 1: from the Channel Tunnel to Fawkham Junction

The route follows the existing railway from the Channel Tunnel to Ashford. East of Ashford there is a 1.6-kilometre (1-mile) long viaduct across the River Stour and the railway line to Canterbury. At Ashford, substantial works were required in a 1.3-kilometre (³/₄-mile) box to modify the Ashford International approaches, construct lines around the station for through trains, allow access to the CTRL for domestic services and realign the existing tracks.

From Ashford the route largely follows the M2 and M20 motorways through Kent. On this length of the CTRL there are two major civil engineering works – the North Downs Tunnel and the Medway Bridge. The former is a single-bore, twin-track tunnel 3.2 kilometres (2 miles) long, 13 metres (43 feet) wide and 10 metres (32 feet) high. It was constructed using roadheaders and excavators, with primary support to the ground provided by a sprayed concrete lining. After excavation, a reinforced-concrete inner lining was placed to carry the long-term

loads. The tunnel was excavated above the existing water table through chalk soils.

The Medway Bridge comprises a new viaduct and bridge structure. At 1,255 metres (4,120 feet) long, it crosses the valley of the tidal River Medway about 80 metres (260 feet) upstream of the existing M2 motorway bridge. Between the two is a further motorway bridge, constructed simultaneously with the CTRL viaduct, to accommodate the motorway widening from two to four lanes in each direction. The two structures were designed to be similar in appearance, with spans and profiles matching the existing viaduct.

The superstructure of the bridge is a prestressed concrete box girder, and the approach viaducts were constructed using incremental push launching methods. The centre section, comprising two balanced cantilevers, was cast in situ to provide a main span of 152 metres (500 feet) – the longest high-speed railway span in Europe.

The substructures are slender in situ concrete piers supported on concrete piles in the chalk, with the viaduct spans at about 40-metre (130-foot) centres. The balanced cantilever sections are supported on two piled concrete islands in the river.

Section 2: from Fawkham Junction to St Pancras

Immediately north-west of Fawkham Junction is Ebbsfleet station, and then the line drops to enter the Thames Tunnel – a pair of 7-metre (23½-foot) diameter bores, 2.5 kilometres (1½ miles) long, with 1 kilometre (½ mile) of cut-and-cover approaches.

North of the Thames the line rises onto the 1.2-kilometre (¾-mile) Thurrock Viaduct – which passes under the Dartford Road Bridge approaches – the 1-kilometre (½-mile) Aveley Viaduct, and the 520-metre (1700-foot) Rainham Viaduct to the entrance of the London Tunnels at Ripple Lane. These are being constructed using Earth Pressure Balance Tunnel Boring Machines (EPB TBMs), assembled in the station box at Stratford. They comprise twin bores, also 7 metres (23½ feet) in diameter, which run for 10 kilometres (6 miles) from Ripple Lane to Stratford station box and a further 7.5 kilometres (4½ miles) from Stratford to King's Cross. Here the line emerges north of King's Cross and St Pancras stations, a link to the NLL allowing through trains to the West Coast Main Line to bypass the terminus.

Environmental works

The CTRL is the first main line to be built in Britain for over a hundred years, and so the extent of environmental works needed to mitigate its impact was an issue that had not arisen previously here on such a project. The importance of environmental works and management has increased enormously in the UK in recent years, and the consideration being given to these aspects is far greater than it was when the major motorways were constructed between 1960 and 1985. Because Stage 1 passes through rural areas of Kent the new railway was aligned to follow the M2 and M20 motorways so as not to intrude into virgin countryside, and the length where it crosses between them was constructed underground in the North Downs Tunnel. Where the route passes near villages, small cut-and-cover tunnels minimize impact on residents. To reduce noise levels extensive softwood barriers up to 5 metres (16½ feet) high were erected; and to contain wheel noise close to the source, galvanised steel panels with absorbent linings on structures 1.4 metres (4½ feet) high were mounted on the track ballast retention kerb.

Extensive archaeological surveys were undertaken, initially using non-invasive resistivity and magnetometer testing techniques. Trial trenches were then dug and 45 significant sites excavated. Five of these were of national importance and include a Roman villa and cemeteries, an Iron Age long house, Anglo-Saxon cemeteries and an Anglo-Saxon watermill.

Much of the route passes through woodland, some of it 'ancient' – which is defined as having been in continuous use as woodland for over 400 years. Where these woods were disturbed, soil was collected and transferred to sites adjacent to other woodland to ensure that the seeds of plant species within the soil could grow in the replacement habitat alongside newly planted native species. Over 1.2 million nursery-grown trees were relocated to 255 hectares (630 acres) of new woodland areas in Stage 1. In addition, 450 hectares (1,112 acres) of species-rich grassland were created using 14 specialist grass and grass/wildflower mixes, and 40 kilometres (25 miles) of new hedgerow planted. Agricultural land used temporarily during construction was fully restored using topsoil carefully collected prior to other works.

Animal species were also affected. New homes were created locally for water voles, badgers, grass snakes and smooth and great crested newts. One hundred hazel dormice were trapped and later released in ancient woodlands in the Midlands as part of a national species re-introduction programme. Monitoring and radio tracking show that they are breeding well in their new environment.

A range of listed buildings of architectural and historic interest was affected by the works – from St Pancras station to medieval timber-framed domestic properties. Among those taken apart piece by piece and re-erected at suitable nearby locations were a 17th century barn, an early 19th century model farm complex, a 15th century Wealden hall house and various cottages. The farm is being reconstructed for educational purposes at Woodchurch Rare Breeds Centre and other buildings are to be located in the Museum of Kent Life. It was considered that the 16th century Bridge House at Merstham would suffer too much damage if dismantled and reconstructed, so it was jacked up on a pre-installed concrete ring-beam foundation and slid 55 metres (180 feet) on steel runners to a new location.

Environmental awareness training was given to all staff at induction and on subsequent occasions, and the observance of environmental policies and commitments was closely monitored.

Conclusion

The UK's first new mainline railway for over a century is being accommodated within areas of outstanding natural beauty, carried over and under major rivers, and threaded through the existing infrastructure of one of the world's principal capital cities. However, the challenges presented by the project were not just of an engineering and environmental nature. To meet the political and financial requirements of the CTRL an entrepreneurial spirit was also required. In rising to all these challenges Arup demonstrated the flair and vision of earlier British railway builders such as Brunel, Stephenson, Hudson, Peto and Brassey. In addition, modern expectations of environmental management, health and safety issues, and public involvement in major projects gave rise to challenges very different from those faced by the 19th century pioneers.

Opposite: Precast elements, such as tunnel linings, were constructed under factory conditions to ensure consistent quality and to avoid disruption to production caused by bad weather.

Left: Various listed buildings of architectural or historic interest were affected by the construction of the CTRL. This 16th-century house was jacked up and moved 55 metres (180 feet).

> St Pancras Station

London, UK, 2001–2007

Summer 2001 saw the start of the major construction programme to realize the vision for a new St Pancras as Britain's international rail gateway to Europe. By 2007, the Grade 1-listed station will have been massively extended and transformed to become the main London terminus for the high-speed Eurostar trains.

The smallest, but most complex, of the three sectors forming Section 2 of the Channel Tunnel Rail Link (CTRL) is Area 100, which comprises the new railways and highways infrastructure over the King's Cross railway lands and the works to and around St Pancras station. This is one of the largest and most challenging development schemes to take place in modern times on a working railway anywhere in the world. There is a wide range of stakeholders involved in the project, including the station owner, London & Continental Railways (LCR); the London Borough of Camden; English Heritage; Network Rail; the train operating companies; London Underground Ltd (LUL); Transport for London; and the statutory utilities. At every stage Rail Link Engineering (RLE) is working in close collaboration with them in order to keep the existing infrastructure in operation whilst carrying out over £600 million of construction.

Opposite: The station, which is being extended to form the London terminus for CTRL, is behind Sir George Gilbert Scott's magnificent hotel building, completed in 1873.

New railway infrastructure

The new St Pancras station will accommodate three separate groups of railway services. The Eurostar trains using the CTRL are about 400 metres (1,300 feet) long, which is twice the length of most conventional trains, including those currently serving the station. In addition to the CTRL connecting to six international platforms there will be grade-separated approaches connecting to three platforms for high-speed domestic commuter services from Kent. The existing Midland Main Line (MML) services to Leicester, Derby, Nottingham and Sheffield will be realigned into four new platforms on the western side of the extended station. New railway connections will be formed both from St Pancras and from the CTRL to the West Coast Main Line via the North London Line (NLL). The existing North London Line connection between the NLL and the East Coast Main Line (ECML) will also be realigned. The Thameslink line carrying cross-London services is currently served by a station to the east of King's Cross. This line will be served by a new station – below Midland Road to the west of the international terminal – and be upgraded from 8 to 24 trains per hour (tph).

The design challenge

The works in this area have to achieve design solutions that will satisfy a Heritage Deed protecting the Grade 1-listed structures of Gilbert Scott's hotel and W. H. Barlow's great trainshed. They include a new underground station concourse, twin tunnels linking to the ECML for the Thameslink 2000 project, and the relocation of cement and aggregate plants in the King's Cross railway lands. The new railway infrastructure and station also require the major redesign and realignment of the local highway network, a major gas distribution complex feeding central London, and the Fleet sewer.

Above: The platform level is elevated to allow trains to cross over Regent's Canal on the approach to the station. The ground level will contain the concourse for the international station.

Opposite: The main entrance to the international terminal is at the side of the station, set back from Euston Road in a new pedestrianized area.

The station

MML platforms will be at the northern end of the old trainshed, entirely under the new canopy, and the MML booking office will move to ground level in the central area at the junction between the existing trainshed and the new station extension. Removing a platform from the trainshed solves the challenge of how to integrate the two levels of the station and hugely enhances the attractiveness of the street-level space by letting daylight into it. Large slots will be cut into the station deck, creating a genuine two-level space where users will see both levels and be able to move between them. The perceived volume of the station will thus be increased from platform level down to street level, transforming the area below the station deck from a liability into a major asset.

The station's new main entrance will be at ground level on the east side. There will be a corresponding major entrance on the western side on Midland Road, and it will still be possible to enter from the south via the forecourt to St Pancras Chambers. St Pancras Road (which at present runs along the eastern side of the station) will become one way northbound and will be diverted to run round the eastern side of the German Gymnasium and Stanley Buildings, thus providing space for the new station entrance and the wider station extension. On the western side of the station, Midland Road will become one way southbound, thus creating a gyratory system around the extended station. This will solve a common problem at major city-centre transport interchanges – taxi access. Here, a dedicated taxi lane will allow taxis to queue while serving the pick-up point in Midland Road, after having set down passengers for departing trains on the Pancras Road side. New bus stops will be located adjacent to the station entrances in Pancras Road and Midland Road.

Passenger circulation

Most pedestrian circulation will be at street level, from where there will be no need to go up or down more than one level: up to the platforms and the platform-level retail outlets and down to the Thameslink platforms. At the Euston Road end of these street-level facilities, beneath the St Pancras Chambers forecourt, the main north-south circulation concourse will lead directly into LUL's new western ticket hall, giving access to the sub-surface lines and linking to the refurbished and extended LUL central ticket hall.

At the northern end of this north-south concourse, the heart of the extended station will be the east-west concourse running across the full width between the entrances and linking via a subway connection to the new LUL northern ticket hall and King's Cross station. Access to MML platforms 1–4 and CTRL domestic platforms 11–13 will be directly from this east-west concourse, as will access down to the new Thameslink station.

Beneath the tracks north of the pedestrian concourses and retail outlets there will be a coach station, complete with group baggage and left-luggage facilities. North of this there is to be a two-storey car park, accessed from the realigned St Pancras Road. This road will then join with the straightened Goods Way to pass under the train deck, forming the public highway link to Midland Road and the continuation northwards of St Pancras Road towards Camden. Finally, north of St Pancras Road, the space beneath the train deck will house a catering centre for the trains, with direct access up to the country ends of all 13 platforms.

The length of Eurostar trains necessitates platforms in excess of 400 metres (1,300 feet) long. The train deck is elevated at St Pancras as a result of the original decision that the track approaches should bridge over the Regent's Canal, rather than pass under it as at King's Cross. This was desirable to prevent the station forming a major barrier in the townscape. Opening up the ground level at St Pancras and making more of its features will assist in achieving this same objective.

For international departures all the facilities will be immediately under the trains, with multiple travelator links up to the platforms. After analysis of the working of the Waterloo terminal, the design team opted for 12-degree-inclined travelators alone, rather than both escalators and travelators. Travelators are far more convenient for passengers with baggage trolleys and baby-buggys (strollers).

Arriving international passengers are dealt with differently. The natural tendency for passengers leaving a train at a city terminus is to walk forwards towards the buffer stops. The time taken for the passengers to walk down the platforms from the various carriages naturally controls the flow through any barriers – in this case immigration and customs controls. So by having long, 6-degree-inclined travelators at the ends of the platforms, it should be possible to avoid large queues at passport control in the Arrivals Hall.

The capacity of the public spaces and vertical passenger movement were tested analytically by sophisticated analysis of pedestrian circulation. As well as designing for safety and capacity, computer modelling was used to establish footfall figures for the optimum location of retail facilities. For international train travel to compete with short-haul flights, St Pancras had to be planned for circulation efficiency as well as for good passenger facilities. It is therefore possible to board a train within 5 minutes of arriving at the international taxi-set-down area.

At peak periods, the station circulation will allow for in excess of 40 trains per hour, with up to three international train departures within 15 minutes, two of which will be separated by only 3 minutes. The international capacity – which is only approximately one-third of the total number expected to use the station – roughly equates to the passenger numbers currently using Heathrow Terminal 4.

The station structure

When completed in 1868, Barlow's trainshed roof was the largest clear-span enclosure ever built. The arched roof will be cleaned, restored, repainted and reglazed. Investigation of the paintwork has shown that the original colour was a dark brown, but that this was soon after replaced by blue – and it is this colour that is to be used again. The glazing to the crown of the arch will be restored to the pattern originally chosen by Barlow, allowing much more light into the space below.

At platform level, the west wall of the trainshed is already pierced by a number of openings, but more will be made, in sympathetic style, to link the platform area directly with the retail area above.

The original intention at the lower level was to retain Barlow's column, girder and plate structure almost in its entirety. It is the ceiling level of this undercroft that provides the tie for the roof arch, and so its integrity has to be preserved. Barlow had been sufficiently far-sighted to realize that platform layouts would change over the lifetime of his station. He had therefore designed the ceiling level of the undercroft as a horizontal deck structure to carry the track beds, with the platforms built up off it. Investigations showed that although the cast-iron columns and their foundations were in excellent shape and perfectly fit to be reused, the strength of the horizontal beams was questionable, in terms both of maximum load-bearing capacity and of expected lifespan. As a result of this analysis it is proposed that a new concrete deck be cast right across the station on top of the existing deck, formed in such a way that its load is transmitted directly onto the columns.

English Heritage is supportive of these changes and sees the revealing of the undercroft areas of the station, the opening up of new vistas in the building and the creation of an even greater sense of space as positive developments.

The new and longer platforms needed at St Pancras are achievable only by extending to the north. Given the importance of the original buildings, the question of which style this extension should adopt became a major issue. There was never, in fact, any particular symmetry between Barlow's great arch and Gilbert Scott's hotel building, so eventually it was agreed that there should be no attempt at pastiche: the addition would be an unashamedly modern structure to the north of Barlow's trainshed. The new extension, covering all 13 platforms, comprises an aluminium-clad louvre-blade and glass roof, giving northern light. Unlike the existing station which is massive and heavy at street level, the extension's lightweight canopy will float clear above the platform deck. It will be carried some 20 metres (65 feet) above street level on minimalist vertical columns on a 30-metre (100-foot) grid, ensuring that for passengers standing under the Barlow arch there will be no sense of feeling 'shut in'. Continuity is achieved by setting the soffit level of the new roof at the exact level of the wind-truss north gable of the existing arch. By making the new lightweight roof float out at this level, the roof will seem almost to disappear when viewed from the old trainshed.

The old and the new will be separated by a glass transept, more than 100 metres (330 feet) across and extending 22.5 metres (74 feet) from the existing end gable to the extension roof. At either end of this will be the new main entrances, also in glass to provide natural light deep into the station. Daylight will percolate down into the LUL subway on the east side and into the descent to the new Thameslink platforms on the west. Passengers using this space will see above them the international trains and will have an end-on view of the north gable of Barlow's work. Looking north they will be able to see through the glazing of the new roof to the sky.

Construction works

Whereas on much of the CTRL project contracts are geographically long with end-on interfaces, in Area 100 everything is on top of everything else. Despite the initial uncertainty about funding, the Government gave authority to safeguard site acquisition and keep preliminary works going so as to enable major construction works to begin in summer 2001. This was necessary to achieve completion by the end of December 2006.

Work on the gas-distribution system is permitted only during the summer, so the gasholders were decommissioned and replaced by line storage in summer 2000 and the gas governor system on the site was replaced during the following summer. Heritage-listed gasholder structures were also dismantled and stored for future re-erection.

On the railway lands a new siding with run-round loop will be provided west of the MML in order to retain the rail-served cement and concrete batching plant. The tenants were originally going to stay put, which would have left them in between the viaducts, but they have now agreed to move into a new, more compact area between the MML and the new chord connection between the NLL and St Pancras.

Station contracts

The enabling contract (no. 135) started during the summer of 2001. This covered the moving of roads, diversion of the utilities, and demolition works to free up the critical eastern side of the station. This is where the first part of the new deck for the extension is being built, on which work began in January 2002. This is the main construction contract. The contractor will first work on the eastern station extension to provide a new interim station by early 2004, before demolishing the existing approach viaducts and west-side buildings. Then the Thameslink box will be built and the deck of the new station completed, together with the new roof and the whole of the new transept area.

The three other major packages for the station are Contracts 108 covering the refurbishment of the Barlow trainshed, 109 covering the architectural fit-out works throughout the whole station and 139 covering building services throughout the station. The interfaces between these packages are complex, but RLE were not confident that a single, large combined package could be procured competitively. The contracts were advertised simultaneously to give the option for combined prequalifications, and this attracted joint-venture bids leading to the award of a combined contract.

Railway works

It is essential to maintain a two-track approach to the station for MML trains at all times. The first package moved the Midland approach lines to the eastern side of the existing formation north of the station, thus releasing the site for the construction of the new aggregate siding. The Midland lines will occupy the decks of the Camley Street and Regent's Canal bridges that have not been in use for some years, thus freeing the bridges' western sides for new construction.

When the west-side work is completed, the approach lines and station throat will be slewed over as far as possible in the other direction and take existing platforms 6 and 7 out of use. This then allows the eastern part of the station extension to be built. When this is completed, the Midland lines will be slewed east again, into an interim station on the new eastern deck extension, which MML will use between early 2004 and early 2006. This will in turn clear the way for the major works on the existing station and on the west side of the layout. In early 2006 the Midland lines will move into their final position and allow MML to take occupation of its new station. The interim station can then be prepared for its final role as the home of the Kent high-speed services and the international station can be completed during the remainder of 2006.

Conclusion

St Pancras Station comprises one of the finest trainsheds of the Victorian era and Gilbert Scott's hotel with its magnificent scale, presence and detailing. Together they are acknowledged as among the most important buildings in London. Their conversion into an international terminus and a station serving inter-city and regional trains presented a challenge of architecture and engineering, made exciting both by the modernity of the new parts and the history of the old. Skills involving both conservation and imagination have gone into adapting the undercroft to form the international concourse and extending the trainshed with a contemporary structure. The end result will be a fitting arrival point in London for passengers from mainland Europe.

Below: The original trainshed will be used by international trains. The new trainshed will accommodate Midland Mainline Servies to the west (left) and Kent Expresses to the east (right) of the international tracks.

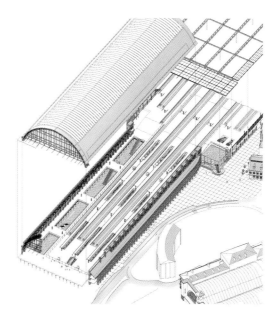

Opposite: The magnificent staircase in the St Pancras Hotel. The building was empty for many years, but the prospect of the International Station has attracted developers who will convert it to provide a modern hotel and apartments.

> King's Cross/St Pancras Underground Station

London, UK, 2004

The first underground railway in the world, the Metropolitan Railway, opened in 1863. It was constructed by the cut-and-cover method between Farringdon and Paddington under the streets of central London, and had intermediate stations serving the mainline stations at King's Cross and Euston. King's Cross – the London terminus of the Great North Eastern Railway, serving York, Leeds, Newcastle and Edinburgh – was opened in 1852. In 1868 St Pancras – the London terminus of the Midland Main Line, serving Leicester, Nottingham, Derby and Sheffield – opened alongside King's Cross. The underground station is an island platform underneath Euston Road, which is part of the central ring road and thus one of the main distributor roads of modern-day London. St Pancras station is at high level, with an access ramp constructed over the eastbound platform of the Metropolitan Line station.

By 1884 the Metropolitan Line had become part of the completed Circle Line which surrounds the central business district of the capital and serves most of the mainline rail termini. Extensions of the Metropolitan Line to Wembley and into the Buckinghamshire countryside, and the later Hammersmith and City Line are also now part of the Metropolitan/Circle Line system and use the same platforms under Euston Road.

Opposite: The isometric shows the complexity of the project. St Pancras mainline station is to the left and King's Cross to the right. Several underground lines serve the stations: the Circle/Metropolitan/Hammersmith and City (yellow), Piccadilly (purple), Victoria (pale blue) and Northern (grey). Thameslink services across London also pass under the stations (as shown by the thick blue line, middle left).

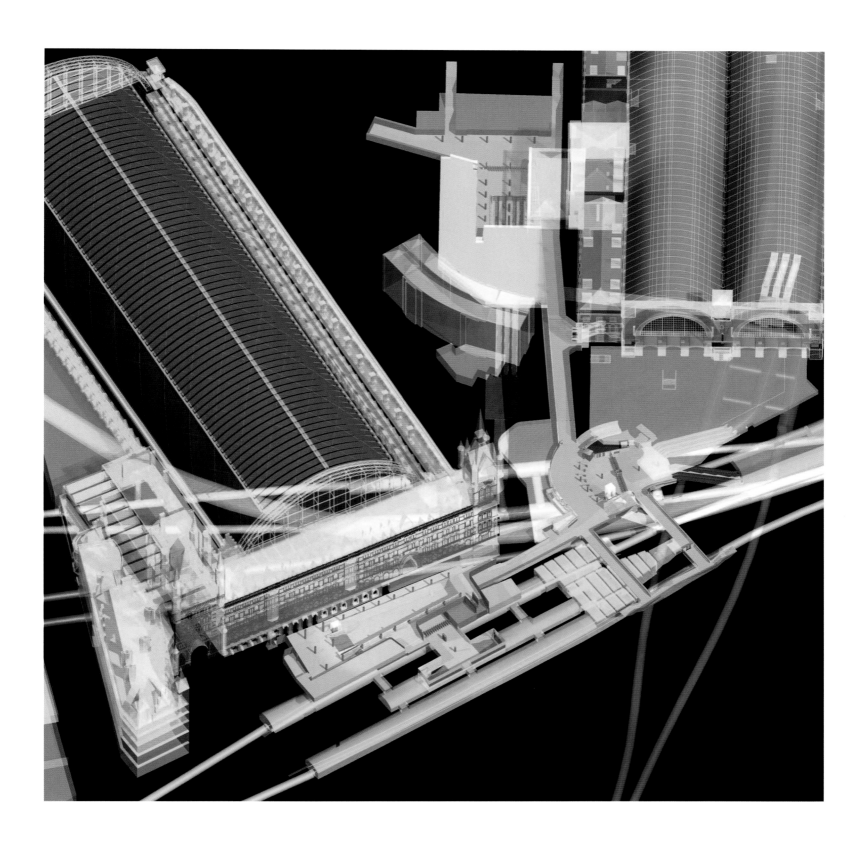

Above: King's Cross Station was part of the first underground railway ever built. This station was to the east of the present site.

Opposite: The entrance from Euston Road uses the arches below the St Pancras approach ramp.

When the deep tube lines were constructed – the Piccadilly in 1906, the Northern in 1907 and the Victoria in 1968 – they were all routed through the important interchange of King's Cross and share a ticket hall and concourse under the station forecourt. This ticket hall has an exit to the pedestrian subway under Euston Road which provides access to the Metropolitan, District, and Hammersmith and City concourse situated between the sub-surface platforms. In London Underground parlance, the early cut-and-cover lines, originally operated by steam traction, are known as the 'sub-surface lines' and the bored tunnel lines, which have always been electrified, the 'deep tube lines'. The underground station is called King's Cross/St Pancras, as it serves both mainline termini.

Some suburban services on the Midland Main Line passed under St Pancras station on their way to Moorgate, stopping at a separate King's Cross station in Pentonville Road. In 1988 this service was extended through the old Snow Hill tunnel under central London to join with services into Blackfriars, creating the cross-London service known as Thameslink. There are now Thameslink through-trains from Bedford and Luton in the north to Sutton, Gatwick Airport and Brighton in the south, passing through the King's Cross Thameslink station. The entrance to this station is in Pentonville Road (the eastern extension of Euston Road) and there is a tunnel linking the station with the low-level Victoria Line concourse. The six tube lines, Thameslink and the two mainline stations make up the largest passenger interchange in London. Many of the links are complex and inconvenient, reflecting the piecemeal development of the individual elements.

Future development

Two new stations are planned. By 2007 the Channel Tunnel Rail Link (CTRL), now under construction, will bring into St Pancras international Eurostar services from Paris and Brussels and express domestic services from Kent, in addition to the Midland Main Line services. St Pancras will effectively become three stations, with the Midland Main Line and Kent Express platforms sharing the upper-level concourse and the Eurostar services using a concourse below the tracks in the former goods storage areas. The Thameslink service is to be upgraded from 8 to 24 trains per hour and will serve a wider range of destinations. The Thameslink station, already overcrowded, will be replaced by a new station under Midland Road, to the east of St Pancras between the station and the new British Library.

The existing London Underground station is one of the busiest on the system, serving 55,000 passengers during the daily morning peak – a figure which is predicted to rise to 82,000 by 2011.

Two new ticket halls are proposed, and the existing underground ticket hall is to be extended with new access subways and passages to improve connections. The route between the existing ticket hall for the deep lines and the Metropolitan Line platforms is currently via a public subway and requires interchange passengers to pass through two ticket barriers. A new public subway to be constructed under Euston Road will do away with this inconvenience and thus make the interchange easier and reduce opportunities for ticket fraud.

Phase 1, generally referred to as the Southern Works, includes the enlargement and refurbishment of the existing underground ticket hall and the construction of a new two-level western ticket hall under the St Pancras forecourt ramp, together with a new interchange under Euston Road. The western ticket hall is to serve the Circle, Metropolitan, and Hammersmith and City tube lines. It will also provide direct access to the concourse of the international station in the undercroft of St Pancras. This phase will also provide step-free access to the Metropolitan and Circle Line platforms for mobility-impaired travellers.

Phase 2, identified as the Northern Works, comprises the new northern ticket hall with cut-and-cover passages to the existing tube ticket hall and the CTRL at St Pancras, and access to King's Cross mainline station. A bank of escalators will lead down to a low-level interchange hub from where access tunnels and escalators will lead to the Northern, Piccadilly and Victoria Line platforms. New lifts to platform level will ensure that mobility-impaired passengers have access to all areas.

Design and construction constraints

Reflecting the history of development in the area are several buildings of historic interest which cannot be altered or demolished without the approval of English Heritage. They include St Pancras and King's Cross stations and, located between these two, the Great Northern Hotel and the German Gymnasium.

The hotel and the offices of St Pancras are of a delightful red-brick composition with myriad turrets and a highly decorated façade designed by Sir George Gilbert Scott. The building has been vacant for many years and is to be restored as a hotel with apartments on the upper levels. King's Cross is more austere, and its historic façade is partially obscured by a meretricious modern ticket hall, but this is the subject of a temporary planning consent and is due to be replaced by a new ticket hall on the western side of the station trainshed. St Pancras and King's Cross stations have the highest grade of listing, Grade I, which is given to only a very small proportion of the buildings considered worthy of the special protection that listing confers. The stations are therefore considered as two of the most important historic buildings in London.

King's Cross Underground station was the scene of a tragic fire in 1987 which killed 31 people when grease and dirt under the wooden steps of an escalator ignited and fire flashed over the roof of the ticket hall, trapping hundreds of people. As a result, very strict safety regulations covering the design and operation of underground stations were introduced in Britain. All new and altered stations have to comply fully with these regulations.

Works cannot be carried out on concourses and platforms while the station is open to the public. Since trains run until after midnight and start again at about 5am, this effectively limits work in these areas to about four 'engineering hours' at night. At King's Cross, London Underground has only two tracks on the underground sections of all lines, so it is not possible to divert services whilst works are taking place. Before the station can reopen hoarding has to be erected around all work areas and all loose materials and tools removed. Similarly, work over the tracks cannot take place while the railway power supply is switched on. Some work is done during weekend closures of the sub-surface line, when train services are replaced by buses, but these opportunities are strictly limited. Closing platforms during the working week while allowing trains to run through without stopping is not permitted, because King's Cross is such a critical interchange on the London Underground system.

As well as the operating railways and the platforms, concourses, connecting passages and shafts of the Underground stations, there are other underground obstructions. Major sewers run under the site, as does the Fleet river, enclosed in a culvert in 1812 when the Regent's Canal was built. On the King's Cross Lands, which was the site of an extensive gas works, the large gas mains are still operational. There are also four railway tunnels, formerly used either for passenger services or the transfer of empty rolling stock, linking various sections of the tunnels currently in use. All these obstacles are in addition to the electricity and telecommunications cables, water and gas mains and sewers found under the streets of any major city.

On the surface, apart from the existing buildings, there are very busy main roads which need to be excavated for the diversion of utilities as well as for new underground works. Changes to the traffic management arrangements sometimes have to be made on a daily basis. It is also necessary to maintain bus stops and taxi ranks close to the stations, as well as emergency access and safe pedestrian and vehicle routes. Traffic flow in Euston Road has become critically important following the Mayor of London's introduction of a £5 (€8, US$8) congestion charge in 2003. Euston Road marks the boundary of the central area and will be used by traffic aiming to avoid the charge. Where possible, the most disruptive works are therefore programmed to be completed before congestion charging starts.

Opposite: The restricted headroom of the new booking hall is relieved by the sculptural form of the roof.

Below: The project will require new links to each Underground platform, with lifts meeting modern requirements for step-free access.

The Southern Works

Phase 1, the Southern Works, commenced in 2001 with the excavation of the ramp forecourt of St Pancras station. This work is particularly sensitive for not only is support required for the station buildings, but the surfaces and retaining wall of the ramp are important parts of the original construction. The greatest risk of damage to the historic buildings came from ground heave as fill material was removed, and the construction method was designed to minimize this risk. The retaining wall is being supported in situ above the booking-hall excavations, and all materials removed from the site are carefully recorded and stored for future reinstatement. Regular meetings are held with English Heritage and the London Borough of Camden (the planning authority) to agree principles and details before submission of consent applications.

In order to provide access to the westbound platform on the south side of Euston Road, the crowns of the arches of the Victorian railway tunnels have had to be flattened to create sufficient headroom for the concourse under Euston Road. Extensive works to divert buried utilities plant have required many changes to the traffic arrangements on Euston Road, and new bus lanes have been introduced along its entire length to limit the effect of increased congestion on the many bus routes.

The Southern Works involve the construction of new pedestrian subways below Euston and St Pancras Roads. The tube ticket hall expansion has necessitated the relocation of the King's Cross taxi rank to the eastern side of the station and the closure of St Pancras Road, with diversion of traffic and some bus services. Removal of the large gas main in St Pancras Road was effected by relocating it in the disused Hotel Curve railway tunnel which passes under the King's Cross mainline station booking hall.

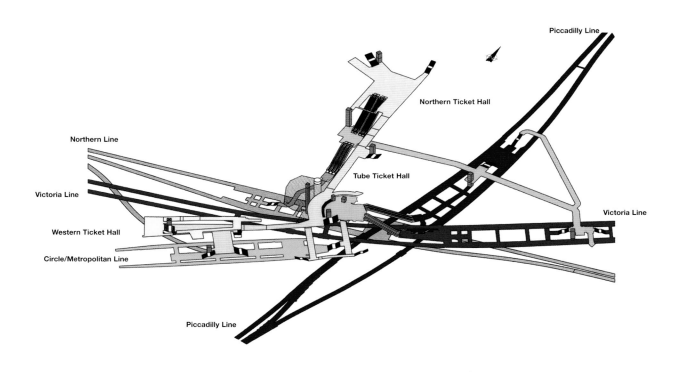

Managing the project

Because the Southern Works and the new CTRL to the north of the main stations are proceeding simultaneously, road closures, diversions and utility works have had to be closely co-ordinated. Meetings are required, sometimes weekly, involving the various parties. Three highway authorities are involved: Transport for London (responsible for Euston and Pentonville Roads), Camden Council (responsible for local roads to the west and south) and Islington Council (responsible for roads to the east). As well as the London Underground ticket hall and CTRL project teams, representatives are present from utility companies, London Buses, the taxi trade, the station operators – London & Continental Properties (St Pancras) and Network Rail (King's Cross) – and other concerned parties.

Conclusion

The original station has been changed many times over its 140-year history. Creating a new Underground station to meet the increased traffic, safety and accessibility demands of the 21st century is a major challenge, for in the centre of a city like London, space underground is almost as crowded as surface space. The solution to the many problems encountered will not be massive architecture or high drama – the site is too difficult for that. However, careful engineering and pragmatic design, together with constant attention to detail, will produce a station that meets current standards of passenger comfort and has the capacity to cater for the new demands. Constantly aware of the tragedy that cost so many lives, members of the design team have endeavoured to create a station that is not only convenient and attractive but one of the safest in the world.

Opposite: The new Concourse is below the St Pancras approach ramp.

Above: Light is used well to reduce the impact of low ceilings.

> Hanging Railway Stations

Wuppertal, Germany, 1993–2003

The Wuppertaler Schwebebahn, built around the end of the 19th century, is a unique public transportation system. It was the brainchild of a highly creative engineer from Cologne, Eugen Lange, whose idea for a monorail suspended from a fixed structure was inspired by the two-rope cable car. The concept was well suited to construction over the River Wupper, with the track supported by inclined steel columns on each bank. Construction of the Schwebebahn started in 1898, the first leg was opened to the public in 1901, and the entire 13.3-kilometre (8$\frac{1}{4}$-mile) length finished in 1903. Since then the monorail, which runs either over the river or through narrow streets, has connected the urban centres of Vohwinkel, Elberfeld and Barmen, and is now recognized as an important landmark in the Wuppertal area.

Lange proposed a monorail instead of a dual-rail system in order to accommodate the large centrifugal forces on vehicles and passengers where the route follows the bends of the river. At these points the monorail's tilt prevents uncomfortable horizontal accelerations. This same tilting train principle was adopted towards the end of the 20th century to allow express passenger trains to travel faster around the curves of existing lines.

The Schwebebahn's rolling stock and systems were upgraded in the 1970s but, apart from the repair of wartime damage, the stations and track structures have remained largely unchanged. In the early 1990s a study undertaken by the owner, the Wuppertaler Stadtwerke AG, forecast a major increase in passenger usage. To cater for this demand there would need to be more frequent trains and the portal frames and bridges supporting the tracks would have to be strengthened to take the increased loads. It was therefore decided to rebuild the entire century-old structure. Because most of the stations were supported directly by the portal frames of the superstructure, these too had to be dismantled.

Opposite: Loher Brücke station: these new stations meet modern requirements, containing elevators for step-free access in addition to well-designed staircases and other features.

In 1993 an architectural competition was held to develop a design for the first station, Zoo/Stadion. It was won by Jaspert & Steffens, with Schuster Architekten as runner up. Together with two other architectural offices – Claudia Drosdowski and Chamier & Molina – they were then commissioned to produce designs to replace most of the existing stations, with Arup GmbH as engineers.

The entire monorail system is a National Monument, and the extent to which the stations should be preserved became the subject of much debate. To retain them all unchanged would have caused major problems in managing the predicted increase in traffic. During the planning process it was eventually decided that the stations at Landgericht and Völklinger Strasse should be preserved, and thus renovated rather than replaced.

Since the Schwebebahn is one of the main means of public transportation in Wuppertal, it was essential that disruption to the system caused by the construction works be kept to the minimum. The main elements of each station – platform, bridges and stairs – were thus finished and available for safe use by the public during a closure of about six weeks, and the remaining fit-out works completed over the following two months.

On the old structure, the trains were noisy and vibration caused disturbance to the passengers and those living close to the Schwebebahn. Moreover, the vibration reduced the durability of the structure and the problem of loosened bolts was a continual maintenance issue. To reduce vibration within the stations, the connections between the cables/tension rods and the columns now have elastomer bearings to provide effective damping capability.

Until reconstruction started, the Schwebebahn had been one of the safest public transportation systems in the world, with very few accidents. Once a derailment had caused a car to fall into the river and once a train struck a cart filled with straw. After a truck struck one of the supports, concrete fenders were added to them all for protection. The most famous accident was when Tuffi, a baby elephant, was taken for a ride during the 75th anniversary celebrations. Tuffi panicked, broke the windows and fell into the river. Fortunately Tuffi was uninjured and has now become something of a mascot for the railway.

The first fatal accident occurred during the construction phase, in April 1999, when a passenger train came off the rails and fell into the river, killing three people. It was found that steelworkers had failed to replace a steel clamp after a weekend possession to replace columns and bridges.

Glass is a strong feature in all the new designs, allowing an open view into the stations and improving security. Most of the old stations had closed façades and many people, especially women, felt unsafe. Lifts at all the new stations facilitate access for mobility-impaired passengers.

The main construction material for all new stations is steel, on foundations of reinforced concrete.

Jaspert & Steffens, Cologne: Pestalozzistrasse, Robert-Daum-Platz, Varresbeck, Westende, Zoo/Stadion

Whereas the platforms of these five stations originally sat on the columns of the portal frames, the new platform structures are suspended from the lintels of the frames by steel cables, and horizontal prestressed tension cables connect with the pillars of the columns to stabilize them. The platform itself consists of a grillage made up of two welded I-sections with a depth of 1,250 millimetres (4 feet). Its horizontal stability is achieved by introducing 114-millimetre (4½-inch) diameter hollow sections for bracing.

Boxes with all the technical equipment required to run the Schwebebahn are suspended beneath the platforms, and are horizontally supported by the bridges. The bridges themselves rest on elastomer bearings supported by new concrete retaining walls on the banks of the Wupper. Since the boxes and the bridges have different vertical supports, relative movements with a magnitude of around 100 millimetres (4 inches) had to be taken into account when designing the connections between them.

The roofs of the stations are directly supported lintels of the portal frames. For all stations except Zoo/Stadion, the roof structure comprises steel trusses with purlins on top.

Opposite left: The Schwebebahn runs over the River Wupper for much of its length. Built in 1901, it is pictured here in 1948.

Opposite right: Decay of the century-old structure required replacement of all the supports.

Below left and right: Jaspert & Steffens chose to suspend the station platform structures on steel cables.

Schuster Architekten, Düsseldorf: Adler Brücke, Landgericht, Loher Brücke, Völklinger Strasse

Suspended construction is again the main principle of the four structures designed by Schuster, with all parts of the stations suspended from the superstructure of the railway arches.

All the stations are divided into different parts in a similar way. Passengers enter at the lower level and ascend via a pair of lifts and staircases to the upper-level platform. The lower level and an enclosed services box are suspended from the main platform level by steel hangers. The platform level itself is made of two main beams, each suspended by a pair of double cable hangers, and with a layer of secondary beams connecting them. The main beams are welded, hollow sections, 1,700 by 350 millimetres (5 feet 7 inches by 1 foot 2 inches); cross-beams are standard HEA700 H-sections. The roof construction is cantilevered with a set of columns and roof beams.

Staircases have an open C-shape with full glazing around the outside. Full-welded hollow sections were used to accommodate the bending and torsional forces within them. At two stations, Schuster Architects adopted the shape of the original staircases. Adler Brücke has a staircase bent around the end of the main beam of the platform, while at Loher Brücke the staircase is turned at right angles to the main beam, making it cantilever from bottom to top. A staircase on the opposite side balances the forces.

Stabilization of the platforms is realized by horizontal connections to the superstructure. Connections to the riverbank stabilize the lower level of Adler Brücke, and at Loher Brücke stabilization is achieved by frame action between staircases and lift shafts.

Claudia Drosdowski, Radevormwald: Oberbarmen, Wupperfeld

The two stations designed by Claudia Drosdowski are structurally very similar to the Schuster stations, apart from the use of a welded I-section for the main beams of the platforms.

Stabilization of the lower levels is attained by frame action between staircases and services boxes. The erection of these cantilevering and hanging structures needed very careful preparation and the precamber had to be accurately calculated.

Fachmarkt

Above: Loher Brücke station: the engineering of the railway is highly visible, with both the structural support and the suspended track in full view from ground level. The columns and cantilevers of the roofs to the platforms reflect this with exposed steel beams.

Above: Hammerstein station: the stations above roads, by Chamier & Molina Architekten, integrate the platform support structure with the track supports.

Opposite: Alder Brücke station: the staircases, external to the platform structure, are fully glazed for weather protection.

Chamier & Molina Architekten, Düsseldorf: Bruch, Hammerstein, Sonnborn

Whereas all the other new stations are above the river, the three stations designed by Chamier & Molina are above roads in narrow streets with houses on either side. The design has columns supporting the rails as part of the structure of the stations. Overall stability is achieved by the use of both longitudinal and transverse frames.

The stairs and lift shafts are not part of the main building but are separate concrete structures fitted in on either side of the road. The platform can be accessed via small bridges over the carriageway. The boxes, which in the design of the other stations are separate structural elements, are here integrated into the staircases.

Conclusion

This unique form of transport has served Wuppertal well for over a century. Despite the coarseness of the structure to modern eyes, the monorail has gained a place in the hearts of the people of the area and transport historians worldwide. With its refurbishment there was inevitably tension between the desire to conserve a piece of history and the need to provide an efficient transport system. If these modifications last another century before further major works are required, then history will probably look kindly on the architects and engineers of the 21st-century Wuppertaler Schwebebahn.

> Light Rail Stations

Hanover, Germany, 1996–2000

Hanover is a city of over half a million people, with over a million in the Greater Hanover region. In 2000 the city, the capital of Lower Saxony, was host to World Expo 2000, the first world fair to take place in Germany. It was planned for 40 million visitors over five months although, disappointingly, only 18 million attended. In order to cater for the expected visitors, a range of city improvements were carried out. About 85 per cent of the city was destroyed by Allied bombs in the Second World War and the city hall, one of the few buildings to survive, was renovated to celebrate the occasion. In addition, motorways were repaired, the airport given a new terminal, the railway station extensively remodelled and a new train station built at the fair site.

Hanover is well known for the Hannover Messe, the world's largest industrial fair. However, the existing area of the fair was not large enough to provide space for all the pavilions required for Expo 2000, and a new area to the east of the existing site was developed.

Opposite: The light rail system in Hanover had to be extended to accommodate the crowds expected for World Expo 2000.

Above: Emslandstrasse station: made with timber.

Below: Kerstingstrasse station: made with fine stainless steel mesh.

Above: Freundallee station: made with red brick.

Below: Krügerskamp station: made with glass.

Above: Feldbuschwende station: made with a mix of metal (plates) and timber.

Below: Kronsberg station: made with pebble cladding.

The new light rail line

The number of visitors forecast for Expo was very substantially higher than the number expected to attend an event such as the Hannover Messe. Crucial for the success of Expo 2000, therefore, was the upgrading of the tram service to a modern light rail public transport system. An essential part of it was the Stadtbahnlinie D-Süd, carrying lines 6 and 16, which connects the centre of Hanover with the new eastern development.

While the world fair was seeking the best design talents to display within the site, the Hanover transit agency, Üstra Hannoversche Verkehrsbetriebe AG, sought to emulate Expo's desire for high standards in the design of its new tram line.

New rolling stock was designed by a London industrial design firm, Jasper Morrison Ltd, complementing the new stations and providing the modern image required. The tram was unveiled at the Hannover Messe in 1997, and was awarded the Transportation Design Prize and the Ecology Award of the German Industry Forum.

The local firm of Despang Architekten won the competition for a total of 13 new stations, with Arup commissioned to carry out the structural engineering for all of them.

Below: The granite platform surfaces rest on a grillage of rolled steel I-sections.

The stations

The design of the stations is based on a modular system that is able to reflect the local environment in each of the neighbourhoods through which the light railway passes. With most of the elements prefabricated, the amount of site work was reduced to a minimum. Prefabrication also improved quality and reduced costs.

Each station has the same four elements:

- a platform
- ramps to provide access for disabled passengers
- stairs
- a waiting area

The side walls of the ramp and the stairs are precast-concrete retaining walls. After they had been placed on site, the gap between the side walls of the ramp was filled with gravel or sand. Concrete for the ramp itself was then cast in situ.

The platform itself consists of two main welded-steel segments made of rolled I-sections. The length of these segments is either 5.5 metres (18 feet) or 4.5 metres (15 feet). The longer segment lies longitudinally on the foundations, which are precast concrete blocks. The shorter element is used transversely between the bearers. To take up thermal movement, the connection between these elements was designed as

Above: Bischofshof/Lange-Feld Strasse: made with a patinated-copper finish.

Above: The cladding reflects the texture of local buildings or landscape.

an expansion joint in a longitudinal direction. This creates a grid. Granite slabs form the surface of the platforms.

The waiting shelters are made of shop-welded steel frames which are rigidly built into the concrete foundations supporting the grid of the platform. Each of the concrete blocks is designed to support a 5.5-metre (18-foot) shelter module if required. The length of the shelter on each platform can therefore be adjusted to accommodate the forecast passenger traffic, or extended in future as needed.

The shelters have vertical panels of 19-millimetre (³/₄-inch) security glass, and there is a glass canopy to provide protection from the elements. The end of the shelter facing oncoming traffic is set back so that passengers can see the trams approaching.

Günther and Martin Despang believe that urban space is not always treated very kindly, so to rectify this they specified various proactive and preventative treatments to deter vandalism: all built-in elements, such as information and advertising boards, fit flush within the structure; finishes are treated with coatings that protect against weather and graffiti; and the construction makes use of smooth, non-adhesive materials.

The cladding of the shelters changes with each station to provide a harmonious relationship with the local urban form. Fixed to prefabricated steel frames are different materials such as steel, copper, glass or hollow-glass blocks, timber or masonry. To keep site work to a minimum, these frames are hung directly onto the steelwork of the shelters.

At the Haltstelle and Freundallee stops, where the prevailing building material in the neighborhood is brick, the facings are of dry-pressed brick. At Bischofshof/Lange-Feld Strasse, the patinated-copper finish oxidizes to reflect the natural colours of the nearby allotments. Kerstingstrasse has a fine stainless-steel-mesh finish, Bünteweg is clad with larch panels in a metal mesh, and pre-cast concrete is the

dominant material at Pressehaus. All the materials were laboratory tested, and even satin-finished glass blocks were found to be sufficiently resistant to abuse to be used to provide texture and variety. In this way, each stop is given a different character – what Despang would describe as 'urban punctuation'. Bold exclamation marks, possibly even question marks, can be encountered along the line before arriving at the full stop of the Messe terminus.

Conclusion

Imaginative use of materials has shown that even structures as humble as tram-stop shelters can provide interest and variety. The different character of each stop enhances the sense of place and the distinctiveness of the urban design, as well as enabling the stop to be readily recognized by regular passengers. The stimulus of Expo 2000 in Hanover not only created a number of large pavilions, but also gave rise to a modest and colourful ancillary infrastructure which remains an inspiration to its citizens in their daily lives. The use of modular construction techniques allowed the costs to be kept to a level commensurate with the scale of the buildings. Architecture does not have to be big to be beautiful.

In 1999 the project received one of the Architectural Review Emerging Architecture awards.

> Chapter 3
Bridges

> Øresund Crossing

Denmark/Sweden, 1993–2000

The Øresund lies between Zealand and the Swedish mainland. The shortest crossing is between Helsingør (Elsinore) in Denmark and Helsingborg in Sweden, 50 kilometres (31 miles) to the north of the route eventually chosen. In 1886 an iron rail and road bridge was proposed across this channel, with the roadway above rail tracks. Another early (1888) proposal for a link across the Øresund, also between Helsingør and Helsingborg, was an 'underwater bridge' carrying a railway in a pipe resting on the seabed. As with the crossing of the English Channel between Britain and France, more than a century was to pass before any fixed link was built.

The Danish islands of Funen and Zealand occupy the area between the Danish Jutland peninsula and the mainland of Sweden. The Danish straits form the sea link between the Baltic and the open sea. The 'Little Belt' between Jutland and Funen was bridged in 1935 and again in 1970, providing seamless road and rail links. A tunnel and bridge link across the 'Great Belt' between Funen and Zealand was started in 1987 and completed in 1998, again providing both road and rail links. Both of these projects were within Denmark and developed and implemented by the Danish Government.

Previous page: The Millennium Bridge, London, was designed as a 'Blade of Light'. The brick building in the background is Tate Modern, converted from a former power station to an art gallery by Herzog & de Meuron.

Opposite: The main towers show the absence of cross-bracing above the deck, which is an unusual feature.

Above: Two temporary props support the main span during construction.

Opposite: The concrete towers for the main cable-stayed span were slip-formed.

The Øresund Crossing links not only the countries of Denmark and Sweden but also the cities of Copenhagen and Malmö, completing a continuous road and rail link around the western side of the Baltic Sea. It was recognized that the development of economic links between the two cities would benefit both countries and create an attractive growth area. The link forms part of the Trans European Networks developed by the European Union to assist the free movement of trade and people. A further major link, between Zealand and Germany across the Fehmarn Belt, is proposed to complete the long-distance fixed links in the western Baltic region.

A treaty signed between the Swedish and Danish Governments in 1991 formed a binding agreement to construct the link. It specified that the link should comprise a dual two-lane motorway and a twin-track high-speed railway built on a line south of the island of Saltholm, a nature reserve in the centre of the channel. There was to be an immersed tube tunnel under the Drogden Channel (between Copenhagen and Saltholm), an artificial transition island south of Saltholm, and a high-level bridge with spans over the Flintrännan and Trindelrännan navigational channels in Swedish waters. Sweden and Denmark were to guarantee the loans, and the construction and operating costs were to be recovered through tolls. The two governments formed an equally-owned joint company, Øresundskonsortiet, responsible for the financing, design, construction and operation of the link.

A design competition was launched, allowing for some variation on the 'reference project'. Of particular importance were environmental factors, for the structures could affect not only the ecology of the straits themselves, but potentially that of the entire Baltic Sea. Changes in the flow through the channel could affect the salinity of the Baltic Sea, and consequently the marine ecology and the amount of ice formed in the approaches to Baltic ports. The two countries established a criterion that water flow through the straits should not be impeded by more than 0.5 per cent.

The successful design, by the Arup-led ASO group, varied from the reference design in several respects. Firstly, the railway was placed underneath the road, not alongside it. In the competition submission this split the central island in two, with the road being carried on a low-level bridge and the railway in a tunnel between the two halves. This reduced interference with water flows, although subsequently the two parts of the islands were combined. The alignment, curved in plan rather than a straight crossing of the main channel, introduced additional visual interest. It also meant that there could be a single link across the Flintrännan, the larger of the two navigational channels. With the railway lines below the road, the depth of truss required was fixed by railway clearances. This allowed for a longer main span on the cable-stayed bridge than that proposed in the reference case. Instead of two bridges with spans of 330 and 290 metres (1,083 and 951 feet) across the two navigation channels, a single span of 490 metres (1,608 feet) could be provided without heavy cost penalties. As well as saving money, this option avoided the visual conflict that two cable-stayed bridges of different dimensions would have created.

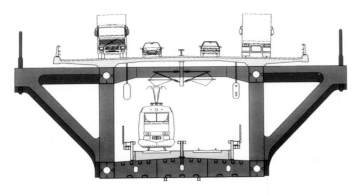

A fundamental principle in the development of the conceptual design was that it should allow for economical construction and have minimal adverse impact on the environment. The objectives of the ASO team were to demonstrate that the design was practical, that it could be built within the implementation programme set by the client and that it would allow for competitive tendering for the proposed design-and-construct contracts. These objectives were achieved by incorporating scope for the following:

– factory prefabrication of large sections of each element
  of the link
– large-scale erection operations for which tenderers would be
  able to use existing plant
– repetition of detail design and construction details

The tender of the selected contractor was based to a large degree on these assumptions. The principles behind the client's contract strategy were:

– detailed design and build
– application of well-known technology
– control and documentation of quality
– division of risks attributable to ground and weather conditions
  and the obtaining of permits

Full responsibility for quality control of the works lay with the contractor. The ASO team worked closely with the owner's staff and were responsible for technical monitoring and audits of the contractor's detailed design work.

The total length of the crossing is 15,840 metres (51,970 feet). This comprises, from west to east, an artificial peninsula (430 metres/1,410 feet), the tunnel (3,510 metres/11,515 feet), the artificial island (4,055 metres/13,304 feet), the western approach bridge (3,014 metres/9,888 feet), the main bridge (1,092 metres/3,583 feet) and the eastern approach bridge (3,739 metres/12,267 feet). The main bridge has a main span of 490 metres (1,608 feet) and two approach spans on each side, one of 141 metres (395 feet) and one of 160 metres (525 feet). The western approach bridge comprises four 120-metre (395-foot) spans and 18 spans of 140 metres (460 feet), and the eastern approach bridge three spans of 120 metres (395 feet) and 24 spans of 140 metres (460 feet). In total, therefore, the crossing has 54 spans and a length of 7,845 metres (25,738 feet).

For the approach bridges, spans were standardized for economical construction. The side spans of the cable-stayed bridge differed in geometry because the inclined members in the trusses were aligned with the cables to carry the load through to the lower deck. At this point the water is shallow, and the piers were founded between 3 and 9 metres (10–30 feet) below datum on the limestone seabed. The limestone was exposed and three pre-positioned concrete pads were used to carry each pier and pylon caisson. Each pair of pylons of the cable-stayed main span is independent above deck level and was constructed in in-situ concrete using a traditional climbing form.

The shallowness of the water, combined with the realignment of the Flintrännan navigational channel, allowed the construction of temporary supports for the main span, which was built in four sections. This allowed all the deck sections to be constructed within the 9,500-tonne (9,350-ton) loading capacity of the heavy-lift barge Svanen. This vessel had also been used to place all the pier caissons, apart from the 19,000-tonne (18,700-ton) caissons for the two pylons, which were placed by a purpose-built catamaran.

The pylon caissons were cast in a dry dock in Malmö Central Harbour; all other caissons came from a fabrication yard in Malmö North Harbour. The steelwork for the approach spans was fabricated at Cadiz in Spain and transported in pairs on ocean-going barges to Malmö North Harbour, where the concrete roadway decks were cast. The cable-stayed deck girder was produced in Karlskrona, Sweden, some 200 kilometres (125 miles) from the site, and transported in 140-metre (460-foot) long sections, and the deck itself was cast in the Malmö North yard. The fabrication of steelwork for major bridges at a location some distance from the site is not uncommon, for the skills and facilities available in established fabrication yards far outweigh the extra transport costs involved.

The completed crossing, opened in 2000, is the longest railway bridge in Europe, and the 490-metre (1,608-foot) cable-stayed span is the longest in the world to carry both road and rail traffic. In the first six months rail traffic was higher than expected, with 2 million passengers carried. Revenue from road traffic was under budget, but various season-ticket and incentive tariffs are being used to boost trade. Building costs were 0.5 per cent under the budget set in 1990 (in real terms), and with interest on the loans lower than forecast, the financial returns look reasonably healthy.

Conclusion

The Øresund Crossing is one of the international *grand projets* of the latter half of the 20th century. The conception and execution of projects such as this require vision and boldness. In this case, these attributes were provided by the Danish and Swedish Governments, who established the financial and organizational structures that enabled the project to be realized. The European construction industry responded to their vision with high-class engineering, fabrication and contracting skills to produce one of the major bridges in the world. Despite the development of telecommunications and aviation, people and goods still need to move across the surface of the planet. Projects such as this crossing bring countries and peoples together, as well as providing a dramatic example of human ingenuity.

Opposite left: The twin-track railway line passes through the deck truss, with the road on the upper flange.

Opposite right: A bridge span being transported into position.

Left: The total length of the bridge structure, including approach bridges, is 7.8 kilometres (5 miles).

> Millennium Bridge

London, UK, 1997–2002

The Thames in London is crossed by many fine road and rail bridges, most of them constructed in the 19th century. Surprisingly, there are no footbridges in the tidal section between Teddington Lock in west London and the sea. All the road bridges have footways; one railway bridge, Hungerford Bridge, has new footbridges partly suspended from it on both sides; and there are two foot tunnels, at Woolwich and Greenwich. No completely new bridges were constructed in London in the 20th century, although Waterloo Bridge was replaced in the 1940s and London Bridge between 1970 and 1973. The Dartford Bridge, which carries M25 orbital traffic across the Thames, was built downstream from London and opened in 1991. The last completely new bridge crossing in London itself was Tower Bridge, completed in 1894.

In September 1996 a competition was organized by the *Financial Times* in association with the London Borough of Southwark to design the first footbridge across the tidal section of the Thames in London. The area specified was between Southwark Bridge and Blackfriars Bridge, but the precise location and orientation were not prescribed in the competition brief.

Opposite: The only new bridge to be constructed across the Thames in the 20th century links the tourist attractions of St Paul's Cathedral to the north and Tate Modern to the south.

Above: The illumination of St Paul's Cathedral and the bridge creates magnificent night-time views.

Opposite: The Y-shaped support columns are designed to withstand impact from shipping.

The south bank of the river at Southwark was predominately industrial, but recent development – including the Financial Times building on the site of a former brewery, the construction of a replica of Shakespeare's Globe Theatre close to its original site and a continuous riverside walk – has opened up the area. The Millennium project conversion of the disused Bankside Power Station into London Tate Modern was correctly forecast to be a major attraction, confirming the location as one of prime interest to tourists. The section of the south bank from Westminster Bridge to Tower Bridge now has a variety of tourist destinations and draws in thousands of visitors every day. The area boasts the London Eye, the Royal Festival Hall, the National Theatre, the Hayward Gallery, the Oxo Tower, the preserved warship *HMS Belfast*, and a variety of modern and historic pubs, as well as Tate Modern and the Globe Theatre.

The bridge competition attracted over 200 entries and was won by a team comprising Arup (engineers), Foster and Partners (architects) and Sir Anthony Caro (sculptor). Their design had a very low profile and was to be lit from a low level on the deck. It was described as the 'Blade of Light' and was intended to be a landmark feature by day and night. The site has St Paul's Cathedral to the north, visible from the river along Peter's Hill, and Tate Modern to the south. The location of these buildings determined the alignment of Peter's Hill as the axis of the bridge, from where there is a clear view of the cathedral. Longer-distance views of St Paul's from key locations in London are covered by special planning controls that prohibit development within defined three-dimensional zones. The bridge therefore had to have a low profile to keep below the protected zone, but it also had to be high enough to provide clearance for river traffic. A cable-stayed or conventional suspension bridge would have encroached on the protected views of St Paul's.

Left: Cross-beams suspended from the cables support the deck.

Above: The flat profile of the support cables.

The design concept

The bridge is an unusually shallow suspension form, with the cables largely below the deck to prevent obstruction of views. Two groups of four 120-millimetre (4³/₄-inch) cables are stretched from bank to bank over two intermediate piers in the river. The northern span is 81 metres (266 feet), the centre span 144 metres (472 feet) and the southern span 108 metres (354 feet). The sag of the cable profile is 2.3 metres (7¹/₂ feet) in the main span, which is about six times shallower than a more conventional suspension bridge structure.

Fabricated steel box sections span between the cable groups every 8 metres (26 feet). The 4-metre (13-foot) wide deck structure comprises two steel edge tubes which span onto the transverse arms. The deck itself is made up of extruded aluminium box sections which span between the edge tubes, to which the handrail and lighting are attached. At the southern end of the bridge the deck bifurcates, with the walkway turning back on itself at the abutment to ramp down between the two sides of the bridge to the walkway level on the bank. In this section the decks are cantilevered off the cables on each side and the edge tubes are strengthened to accommodate the torsional loads.

The piers are of reinforced concrete, and the foundations of each one comprise a pair of 6-metre (20-foot) diameter caissons, dug to 18 metres (59 feet) below riverbed level and connected by a 3-metre (10-foot) deep pile cap. Steel V-shaped brackets are mounted on the piers to support the cables. The piers had to be designed to resist impact from any of the many vessels that use the Thames. A survey showed that 400 pass by this location every day. Many are recreational and sightseeing craft, but there are also commercial vessels carrying domestic and industrial waste, oil products and sand and gravel. The largest vessel in use at the time of the analysis was the *Tracy Bennett* (55 metres/180 feet long and 1,210 dead weight tonnage).

Model studies were undertaken to assess the impact of the new bridge piers on the riverbed, and permanent bed protection was provided to prevent scour. These checks also showed that turbulence around the piers was negligible and would therefore not cause navigation problems.

The stability of the bridge deck was examined using wind tunnel tests on one-sixteenth scale sectional models to confirm the computer analysis. This covered the impact of various types of wind load, in particular the maximum design wind speed for global structural stability and the buffeting effect of gusts of wind at more typical speeds with relation to the comfort of bridge users. The response of the bridge due to the dynamic vertical loads from pedestrians was also examined, taking into account the effect of the inclined cables and the consequent coupling of lateral and torsional modes of vibration.

Above: Two types of damper were installed discreetly to prevent the famous 'wobble'.

Opening of the bridge

Construction began on site with archaeological excavations at the end of 1998. The main works started in April 1999 and the bridge was opened on 10 June 2000.

The opening was celebrated with a mass charity walk over the bridges of London. It is estimated that between 80,000 and 100,000 people crossed the bridge during the first day. Analysis of the video footage showed a maximum of 2,000 people on the bridge at any one time, with a maximum density of 1.3–1.5 per square metre (12–14 per 100 square feet).

Unexpected movements occurred – predominately lateral and mainly on the south and central spans. They were not continuous but occurred when a large number of people were on the bridge, and they died down when the number was reduced or when people stopped walking. The movement caused a significant number of pedestrians to have difficulty walking and they had to hold on to the balustrades for support. Although these episodes were alarming, no injuries were sustained and the integrity of the bridge was not threatened. There was no problem with vertical movements, and in this respect the bridge behaved as predicted in the design studies.

On the following day the number of pedestrians allowed on the bridge was restricted, and the movements occurred only occasionally. The next day, however, the decision was made to close the bridge as a precaution and in order to investigate fully the cause of the movements.

Obviously there was intense media interest, given the large number of people involved and the high-profile location of the bridge. As a result the headline writers called it the 'Wobbly Bridge' – and this is probably the name by which it will always be known to Londoners. Ill-informed comment assumed that the engineers had made an elementary mistake in failing to design for vertical harmonic motion, which is a well-known phenomenon and is the reason that soldiers are usually instructed to break step when marching across bridges. In fact the oscillations in the Millennium Bridge were lateral, not vertical, and provision for such conditions did not exist in any bridge codes. Regardless of the design codes, problems of vertical oscillations caused by pedestrian or wind loads are well known to every bridge engineer. They would never have been ignored, either by the experienced Arup design team or by the checking engineers from Mott MacDonald who carried out the independent check required by UK bridge-design regulations.

A more interesting consideration was whether the problem arose from any of the innovative features of the design. This aroused the attention of the technical press, and the resulting publicity led engineers across the world to report examples of similar phenomena that had previously not entered the mainstream technical literature. The attention that civil and structural engineers consequently received was enormous, and some of the Arup engineers concerned became so well known that they would be stopped in the street and asked if they were 'the Wobbly Bridge man'. There has probably never before been such widespread public awareness of the bridge-engineering profession, except in situations where some fatal disaster or major collapse has occurred.

Opposite: The original opening was celebrated with a firework display and a mass charity walk. The reopening was less flamboyant.

Above: Hundreds of volunteer members of Arup staff were used to load the bridge in order to assess the effectiveness of damping proposals.

Above left and right: Absolute precision was the order of the day when the dampers were installed under the deck.

Opposite: The foundations of the piers were 18 metres (59 feet) below river bed level.

Investigations

The phenomenon discovered on the bridge became known as Synchronous Lateral Excitation. It is caused by a lateral loading effect that develops when groups of pedestrians start walking in step. The chance correlation of a crowd of people's footsteps as they walk over a bridge generates slight sideways movements of the structure. It then becomes more comfortable for people to walk in synchronization with the bridge movement. This instinctive behaviour ensures that the sideways forces they exert match the resonant frequency of the bridge, but their timing increases its motion. As the amplitude of the motion increases, the lateral forces imparted by each individual also increase, as does the tendency to walk in step.

Movements of this type were not covered by any bridge design codes and were not commonly known in the engineering profession. All descriptions of the phenomenon discovered by the design team were examined, but none gave any measurements for the force attributed to pedestrians or explained the relationship between the force exerted and the movement of the bridge deck.

There were three priorities in the investigations:

- to compare the built structure to the analytical predictions
- to measure the force exerted on the structure
- to design a system to reduce these movements to acceptable levels

Tests showed that the analysis of the bridge at design stage accurately reflected the built structure. The problem thus derived from an unexpected loading condition rather than a weakness in the analytical model. This meant that the model could be used for future testing of options for restricting the movement.

A programme of research was undertaken to discover the forces involved so that a method could be established to eliminate the problem. Laboratory tests involving pedestrians walking over moving surfaces were carried out at the University of Southampton and Imperial College at the University of London. It was also felt that the only way to replicate the precise conditions of the bridge was to carry out crowd tests on the bridge deck itself. Tests with 100 people in July 2000 were used to devise a load model. A second series was carried out in December 2000 to further validate the design and to load test a prototype damper installation. This test involved 275 people and demonstrated both that the load model was valid and that the damper installation successfully reduced the movements to acceptable levels.

These results showed that the movement phenomenon was not related to any of the bridge's technical innovations. The same effect could occur on other bridges with a resonant frequency in the same range if they were loaded with a sufficient number of pedestrians.

Other cases of Synchronous Lateral Excitation:

The widespread attention that the Millennium Bridge attracted brought to light similar movement problems on bridges around the word. Of particular note are three other locations where Synchronous Lateral Excitation occurred.

Link Bridge, National Exhibition Centre, Birmingham, UK

Built in 1978, this 45-metre (150-foot) footbridge links the Birmingham International railway station to the National Exhibition Centre. Sideways movement on the bridge, a steel truss on unbraced vertical columns, had been experienced when large crowds crossed over it after pop concerts and other events in the Centre.

Groves Suspension Bridge, Chester, UK

This footbridge – originally built in 1923 but subsequently rebuilt – is 100 metres (330 feet) long and of steel construction with a timber deck. Sideways movement was experienced when large crowds gathered on the bridge to view a regatta to celebrate the Silver Jubilee of Queen Elizabeth II in 1977.

Auckland Harbour Bridge, New Zealand

In 1975 excitation occurred on the northern section of this bridge, a steel box girder with a 190-metre (620-foot) span and two lanes on either side added in 1965. The phenomenon was observed during a Maori demonstration when two lanes were closed and between 2,000 and 4,000 demonstrators crossed the bridge.

None of these cases had been fully researched or analyzed and the experience was not widely disseminated within the engineering profession. It is notable that these are conventional structures of widely differing types – further evidence that the problem on the Millennium Bridge could not have been solely the result of its innovative design.

Design modifications

To stiffen the central span sufficiently to raise the resonant frequency to a level which would prevent the phenomenon occurring, the stiffness of the deck would need to be increased tenfold. This was not possible without adding enormously to the mass, seriously compromising the design concept and involving great expense. It was therefore decided to use a damping system to absorb the energy from the pedestrians and prevent the phenomenon from developing.

The objectives governing the selection of dampers were:

- achievement of adequate damping levels to limit the amplitude of oscillations to a level that caused no discomfort to pedestrians
- limitation of additional weight on the bridge
- durability, low maintenance and resistance to fatigue
- minimal effect on the visual appearance of the bridge

Fluid viscous dampers were selected to control the lateral excitation, similar in principle of operation to the shock absorbers in a car's

suspension. Because the movements were so small the devices were required to provide damping at amplitudes of less than 0.5 millimetres ($^3/_{16}$ inch) displacement. A precision-engineered design, developed for use in spacecraft, was adopted. It was decided as a precaution to introduce also some damping of vertical movement, and tuned mass dampers were selected for this purpose. The majority of the viscous dampers are contained in diagonal members placed beneath the deck; others are located in a visible position as diagonals between the cables and the deck either side of the piers and on each side of the approach ramp on the south abutment. The tuned mass dampers are located on compression springs on the top of the transverse arms beneath the deck. The visual effect on the completed structure is minimal and would not normally be observed by the public.

The results of the research have been widely published by Arup and are being incorporated into bridge design codes. The company is currently advising various bridge authorities on the subject of Synchronous Lateral Excitation.

The modifications were made and the bridge tested by 200 volunteers on 30 January 2002. Following analysis of this test the bridge was re-opened, with minimal ceremony, on 22 February 2002.

Conclusion

The Millennium Bridge was conceived as a dramatic addition to the River Thames. Its design met in full the requirements for an iconic structure representing the best of modern technology and materials. Unsurprisingly when the bridge's problems became apparent, most of the blame was directed at Arup and their innovative design, and they came under intense pressure to find a solution. The very public nature of the closure of the bridge meant that Arup had to respond whilst in the full glare of publicity and criticism from both professional colleagues and the public at large. They did so with determination and a high degree of professional skill, producing solutions that have solved the problem and not compromised the essential features of a stunning design.

> Pero's Bridge

Bristol, UK, 1993–1998

In order to improve the port of Bristol, St Augustine's Reach in the centre of the city was dug in 1248 to divert the waters of the River Frome, a tributary of the River Avon which joined it at this point. The river in Bristol has a very high tidal range, and in 1803 the Avon was diverted and its original course dammed to create the Floating Harbour, which maintained ships at a constant level while they were being loaded and unloaded. The working dock was a significant port for Britain's trade with both America and the colonies, but since the Second World War there has been a reduction in trade with port activities moving downstream to Avonmouth. In the 1970s the commercial port closed, but the Floating Harbour is still used for pleasure craft. Many of the dockside buildings are now being converted to other uses and the dock edge has become a popular route for walkers. The reach, however, was seen as a barrier to movement across the city, especially in view of the activity that could be attracted to the dock area following the regeneration of the former gas works on Canon's Marsh to the west. The city council determined that a bridge should be provided across the reach to connect the area to the city centre and provide a circular walk. The construction of the bridge has made possible the concept of a 'Brunel Mile' – from Temple Meads station in the east to the *SS Great Britain* in the west – and this idea is gaining support in the city.

Arup was appointed to provide engineering design of the bridge, which was funded by contributions from property development in the area. A decision was then taken that the design team should include an artist who would develop an appropriate solution for the crossing to enhance the engineers' design. Arup prepared a design brief in which the essential parameters of the crossing – width, height above the water, navigation width, gradients, handrail specification, and so on – were set out. This brief was issued to a shortlist of artists, who then submitted initial concepts. The selection competition concept submitted by the Irish artist, Eilis O'Connell, indicated an organic form in which the whole bridge was treated as a piece of sculpture, with the crossing experience enhanced by the provision of an observation space off the main thoroughfare. Eilis was appointed on the basis of this approach. Arup decided that the artist should be given free reign to develop her concept, constrained only by what was achievable within the engineering requirements of a public footbridge.

Opposite: The footbridge is curved both in plan and elevation.

The design that emerged is the result of close co-operation between artist and engineers, in which the artist led the development of the design. Its dominant elements – the curve in plan, the rolling bascule operation and the counterweights to the bascule – are all as determined by the artist. Similarly, the choice of cast iron for the deck was a reference to the historic use of this material in the dock setting. There has been much debate about the references for the 'horn' form of the counterweights. For some, they call to mind a ship's speaking tubes; for others, sails or flowers. In truth, they are simply an aesthetic form which provides mass and height, designed to enhance the view from the centre and create a feature of visual interest.

Although the bridge was planned to provide a strategic pedestrian link, it crosses a 54-metre (177-foot) navigable reach of the harbour and has to allow shipping to pass. There is sufficient clearance for the small, regular ferry to pass underneath but for tall-masted vessels, such as those that take part in the Bristol annual regatta, a section of the bridge was required to lift or move aside to give unlimited headroom in a defined 9-metre (30-foot) wide channel.

Below: The horns provide a counterweight to the lifting span, which allows high-masted craft to berth at the historic dockside in the centre of Bristol.

Above: The counterweight principle is commonly used in lifting bridges. Bascule bridges like this are common in the Netherlands.

The bridge has a 3-metre (10-foot) wide footway, with handrails positioned for both disabled and able-bodied persons. Gradients were limited to 5 per cent to ensure that the bridge was accessible for wheelchair users. The basic structure is a central spine with transverse purlins tied with the edge toe rail, which also supports the deck plates. The lifting section is designed with box-section side beams curved to provide the rolling arm of the bascule. The horns are welded steel plate on a structural cage. The two reinforced concrete supports in the water are clad in stainless steel to provide a long-lasting and consistent appearance. Lighting is attached to the balustrade to flood the deck and to provide some security lighting for pedestrians. Viewed from afar, the lights appear as a string of white light across the reach. Uplighters in one central pier illuminate the horns.

Construction was carried out by a local boat yard, thus making available not only a workshop adjacent to the water in the city centre and skills in handling irregular shapes over water for installation, but also, most importantly, skills in forming double curvature in steel plate. The bridge was constructed in the yard in three sections. A trial assembly proved the fit and the mesh of the teeth controlling the rolling action. Factory assembly also allowed for the paint finish to be applied in sheltered conditions. Each section was floated up the harbour on pontoons and lifted from a barge-mounted crane onto prepared foundations.

The people of Bristol were invited to suggest a name for the bridge, and over 50 entries were received. The chosen name, Pero's Bridge, commemorates an enslaved man who came to the city from Nevis Island in the Caribbean in 1783, when Bristol was the main centre of the slave trade in Britain. The naming of the bridge makes an interesting counterpoint to the sculptures depicting slave owners displayed elsewhere in the city. One recent example is a statue in nearby Millennium Square of William Penn, the founder of Pennsylvania and keeper of a retinue of slaves.

Conclusion

This quirky little footbridge shows how a simple structure crossing a short gap can create life and interest in an area popular with both visitors and locals. Being close to the heart of the city, this distinctive feature will be seen by many as they pass by. Modest it may be, but it is a distinct asset to the city of Bristol and a symbol of the regeneration taking place in the former dockland area.

Above: The curves and asymmetry of the horns add to the charm of the bridge.

> Spencer Street Footbridge

Melbourne, Australia, 1997–1998

In 1997 Melbourne Exhibition Centre and Melbourne Convention Centre became one entity – the MECC. The Exhibition Centre, opened in 1996, was designed by Denton Corker Marshall Architects and has 30,000 square metres (320,000 square feet) of exhibition space. The Convention Centre can accommodate up to 2,500 people in both the main theatre and the ballroom. To integrate the facilities of the two venues, the MECC commissioned a new covered footbridge spanning the Yarra River, as well as significant enhancements to the façade and lobby of the Convention Centre, thus greatly improving its ability to host major events. In the process, a public promenade was added to the already lively river precinct, which also contains the Casino and Batman Park.

The new footbridge is both public and urban in its form and manners. Its exaggerated, ribbed form cuts a striking profile against river and sky. As a 'found object', this could be the upturned hull of an unfinished boat or the skeletal backbone of some ancient marine creature.

The semi-enclosed bridge lies parallel with the existing Spencer Street road bridge, interlinking with it at the one-third span points. The striking, cantilevered glass-and-steel wall and roof divide the bridge into two lanes – a leeward, protected lane and an outer, exposed lane. In this way the bridge's unusual design offers protection to convention delegates crossing in inclement weather but also allows unfettered views of the river on the open side.

Opposite: The bridge links the Exhibition and Convention Centres, alongside a major road bridge.

Design features

The southern end connects onto Southbank Promenade under a large, bladed entry canopy to the Exhibition Centre. At its northern end, the footbridge links into the Convention Centre under an enormous wall of translucent green glass which hovers over the Spencer Street footpath and gives the impression of being suspended in space.

Community consultation

The commission for this project included writing the brief, which was undertaken in consultation with the client and the various end-users, including convention and exhibition organizers. An extensive consultative process involved some 15 statutory government agencies, departments, neighbouring landowners and the statutory authorities, for any project on the Yarra River entails a complex overlap of various authority and permit requirements.

The final configuration of the footbridge took into consideration community interests as well as those of the client. For example, the open ends allow for unimpeded pedestrian and cycle access along the promenades on both sides of the river.

Engineering design

For the engineering on this project, Arup was required to think laterally, in order to design structures that would be both functional and economical while contributing to and enhancing the architectural concept. A key feature of the process was the pooling of ideas from team members of all disciplines to produce an integrated design.

One of the biggest challenges was the interface between glass and steel, where all the connections are fully exposed. Traditionally, structural steel and glass are fabricated to very different tolerances and are designed to move in different ways. In both the footbridge and the loggia over the Convention Centre entrance, these two elements are combined. The loggia wall of glass appears as a flat sheet. Its only support is a series of recessed spherical fixings, hidden in the depth of the glass.

Central to the footbridge is a large, raking glass wall which leans into the prevailing wind and spans 3.8 metres (12½ feet) from its base to the cantilevered roof fins. Views along the river corridor are framed by the superstructure, which sits like a sculptural curtain hovering on the horizon.

To achieve the clearest view of the river the glass had to appear unsupported on all its edges. This was done by fixing the base into the bridge deck while pinning the top at two points. The bearings at the top fixings not only restrain the glass, but also allow it to rotate under wind loads, letting it move with the bridge deck rather than the roof.

The bridge structure is formed from a series of warped steel trusses which span the Yarra in three arches, mirroring the profile of the Spencer Street road bridge beyond. Detailed analysis of the way the footbridge works was undertaken to ensure that it did not become unstable under any load condition.

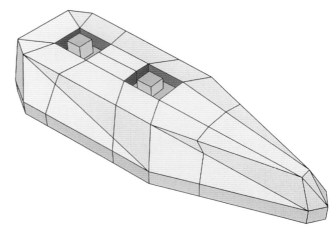

Above: A concrete pier protector shroud in the casting yard.

Above: The profile of the shroud reduces turbulence and the nose is designed to shear off water on impact to protect the pier.

While the base trusses for the bridge warp, the roof canopy cantilevers from the deck and is supported by a series of profiled box sections. These were designed both to handle the structural requirements and to house all the bridge's services. There is no applied cladding so the painted steelwork provides the final finish.

Much effort was put into the design of the cantilevered roof fins to achieve the slimmest fins possible. They were cut from a standard 40-millimetre (1½-inch) plate, keeping wastage to a minimum. A small tube used to prevent them buckling also doubled as the point to support the bearings picking up the top of the wall glazing.

Just as much care was taken with the design of the underside of the structure, given that this is the part seen by people passing below the bridge on the river. This is the best place to appreciate the effect of the stainless-steel props that are pinned to the bottom chord of the truss and rake upwards to form the sculpted handrail.

The installation of the bridge was a major consideration in the project, particularly as it is in Melbourne's central business district. The design enabled the main structure to be fabricated in three pieces, brought to site on trucks and installed in a day. To this base the raking props were added, and the fins forming the canopy bolted to the top in sections.

The footbridge is supported on two river piers sitting on a basalt outcrop. The soil in the riverbed above the basalt is like a thick sludge, so gives no lateral restraint to the piles forming the piers. This became a problem when designing for accidental impact load on the pier and/or bridge, for there is insufficient capacity in the piles to resist this type of load without expensive foundation solutions.

To reduce impact forces, a system of concrete shrouds over fenders on the pile caps was devised. The key design criterion employed was a fully laden dredging barge hitting the bridge and/or shroud at 10 knots (double the speed limit on the river). By hiding fenders beneath the shrouds the bridge's slender appearance was uncompromised. The 'nose' of the shroud is designed to shear under exceptionally high impact, further protecting the bridge.

Loggia/Convention Centre entrance

The Convention Centre glass loggia forms a dramatic new arrival space. Here the artist James Clayden has added a second glass wall as part of the overall urban composition of footbridge and entrance. Shimmering watermark patterns, based on photographs of the rippling surface of the Yarra, have been silk-screened onto the glass. The effect plays tricks with light and surface, adding visual texture on a vast scale.

The loggia's glass screen was made to 'float' above pedestrians in the new forecourt by sitting its entire 12-metre (40-foot) height on four raking, cantilevered columns. A fifth column nearer the footbridge end was omitted to improve circulation. As a result the whole south end of the wall hangs from the existing Convention Centre stair shaft. One mullion thus carries a much greater load than the others, but its careful engineering as a series of welded plates – instead of the boxed channels used for the other members – ensures visual continuity.

The footbridge was opened on time and under budget at the end of 1998. Its success is obvious from the number of pedestrians who choose to use this crossing rather than the road bridge.

Conclusion

The bridge was required to perform two very different functions. Firstly, as part of the city transport network, it had to provide for pedestrian and cycle traffic; secondly, it was to form a link between the Exhibition and Conference Centres. In addition, its exposed and prominent position required a dramatic design, but one sympathetic to the local environment. The bridge successfully meets all these requirements and is thus a striking example of a modest structure that adds value to the landscape as a work of art as well as fulfilling its practical roles.

Below: The view across the river from the loggia of the Convention Centre.

> Denver Millennium Bridge

Colorado, USA, 1999–2002

The city of Denver evolved from five mining settlements near the confluence of the South Platte river and Cherry Creek, in what is today the Central Platte Valley. The land was on the flood plain, development had therefore taken place on the higher ground, leaving in the heart of the city an industrial area and a large railway yard. As part of the recent regeneration of the downtown area nearby, historic districts were redeveloped and an initiative launched to reunite the city, including the construction of the Millennium Bridge. Railway land was purchased and the majority of the tracks removed, leaving just three running through the area on the Consolidated Main Line, in a narrow zone bisecting the new development. The new bridge, linking across the tracks and located in the heart of the new residential and commercial zone, is thus a central marker and icon for both the development and the whole of Denver.

The pedestrian mall on 16th Street was originally designed by I. M. Pei and the proposal was to extend it from Lower Downtown (Lodo) district across the Platte River Valley to the neighbourhoods of west Denver. The mall is tree lined and pedestrianized, with the only vehicular traffic being a shuttle bus system. The extension of this street was key to creating the link across the valley and extending the successful commercial area westward.

Redevelopment included the corridor for a new rail system and a 12-hectare (30-acre) urban park as well as the Millennium Bridge. The bridge spans 40 metres (130 feet) and, in order to continue the theme of the plaza area, is 24 metres (80 feet) wide rather than a narrow linear crossing.

Opposite: The inclined pylon is a landmark feature in the downtown area of Denver.

Structural design

It was decided to minimize the depth of the structure to limit the height pedestrians would have to climb. The provision of a major sculptural element, defining the area and the bridge, was also considered appropriate. The architects, ArchitectureDenver, chose an asymmetrical, single-mast, cable-stayed structure. The 61-metre (200-foot) high mast, which is the main feature of the structure, has a tapered, tubular steel form and is located on the city side of the bridge to one side of the 16th Street thoroughfare. It is inclined towards the city and away from the centre line of the street. Not only does a very tall mast create a defining architectural feature, it also makes the most efficient use of the cables for strength and stiffness.

Steel deck members are suspended on twenty 65-millimetre (2$\frac{1}{2}$-inch) diameter cables connected to the top part of the mast. Five 110-millimetre (4$\frac{1}{4}$-inch) diameter cables anchor the mast to the ground. The back-stay anchors are located along a circular arc defined by a concrete berm which protects people from the lines of the nearby light railway. The resulting twist in the back-stays is an interesting sculptural form and adds to the aesthetic appearance of the structure.

The deck of the bridge is raised 7.5 metres (25 feet) above the surrounding ground level in order to cross the railroads. Staircases and bicycle ramps from both the city and the park side provide the primary access to the deck. For the mobility-impaired, glass lifts provide access from both sides, connecting to the bridge via wooden walkways.

The steel deck members are arranged to form a grillage. Longitudinal members form a fan shape, with transverse members lying across these, aligning with the railroad tracks below. The outermost primary girders are exposed on each side of the bridge deck and form a visually attractive counterbalance.

On the walking surface of the deck the grillage is filled with a sealed concrete carpet. This surface does not completely cover the grillage, but allows the upper surfaces of the structural members to be seen. Large planters and benches help to enforce a sense of destination on this floating piazza, and sculpted connection sockets, where the cables terminate on the deck and at the back-stays, further enhance the bridge's appearance and the pleasure of its users.

The optimization of the cable tensions to balance the deck forces was a huge analytical task. The nature of the structure led to an extremely complex interaction between all the cables, with no direct solution possible. The aim was to maximize the use of the bending capacity of the deck by stressing the cables against it. This positions the bending-moment range to make best use of the hogging and sagging deck capacities at different stages of construction and under different loading cases. Each cable also had to account for a 10 per cent force variation to give the contractor some tolerance.

Opposite left: The bridge is part of a pedestrian route linking the regeneration area to the Central Business District.

Opposite right: The erection of the pylon.

Above: Lighting ensures that the bridge pylon is a focal point by night as well as day.

Right: The structural concept has a backstay to brace the pylon against the support cables.

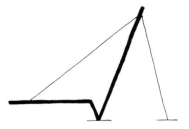

Erection sequence

One of the most challenging aspects of the bridge was its erection. The main girders were placed over the railroad on a line of temporary supports at mid-span point. The secondary supporting beams were then attached. Once the deck members were in place, the mast, which was shipped to site in one piece from the fabricator, was erected and the deck cables attached. With the mast tilted forwards from its final position, the deck cables were attached to the deck fins, having already been cut and adjusted to the desired lengths. The five back-stay cables were then jacked backwards one at a time, moving the mast into its final position. Nine separate jacking operations were required to achieve the correct force in each cable without overstressing any component of the bridge in the process. At each stage the cable forces were monitored and minor adjustments made to ensure they were on target. Finally, the metal deck was placed and the lightweight 165-millimetre (6$\frac{1}{2}$-inch) thick concrete deck poured.

Wellington Webb, mayor of Denver, opened the bridge to the public on 22 April 2002.

Conclusion

The sculptural form of this bridge resulted only partially from the site requirement and the brief. It was also heavily influenced by the desire to produce an icon to represent the urban regeneration taking place in the Platte River Valley. This has been achieved with an elegance of structure and architecture that creates a sense of place on the bridge as well as in the surrounding area.

Below: A back-stay anchorage detail.

Right: A classical staircase provides a complementary feature.

> Corporation Street Footbridge

Manchester, UK, 1998–1999

In June 1996 a bomb placed by the Irish Republican Army (IRA) caused extensive damage to the shopping centre in the heart of Manchester. It was the largest bomb ever exploded by the IRA on mainland Britain, and approximately 20,000 square metres (200,000 square feet) of retail space and 30,000 square metres (300,000 square feet) of office space were destroyed. A new masterplan was drawn up for reconstruction and improvement of the city centre, providing new leisure and cultural facilities as well as retail and office development. A replacement footbridge was planned over Corporation Street, to link Marks & Spencer and the Arndale Centre, two of the city's main shopping locations. Manchester's traffic pattern was also changed by building a new inner ring road and banning traffic other than buses through the centre, thus removing most vehicles from the heart of the city.

The Marks & Spencer store needed to be demolished and totally rebuilt, and the company decided to build the largest in their chain. The Arndale Centre, a large enclosed shopping mall, was extensively damaged, and its refurbishment provided an opportunity to effect considerable improvements to this ugly 1960s building.

As a result of reconstruction work, the centre of gravity of the downtown area shifted slightly westward, to Corporation Street. An international design competition was held by Manchester City Council for the new footbridge across the street. It forms an enclosed link between the new Marks & Spencer flagship store and the refurbished Arndale Centre.

Opposite: The deck of the footbridge is enclosed to protect shoppers from the weather – Manchester is renowned as the rain capital of Britain.

The bridge has set new standards of structural and façade engineering in urban footbridge design. Innovative techniques used in its realization range from the use of prestressed steel to form the hyperbolic paraboloid of the primary structure, to the incorporation of state-of-the-art glazing materials to ensure public safety in the event of glass-panel failure.

One of the biggest design challenges set by the brief was to conceive a form that could both accommodate a sloping pedestrian route and address the street sympathetically. To add to the geometrical complexity was the fact that the openings in the two buildings were not exactly opposite each other and their façades were not quite parallel. However, the structural form of the bridge, together with the fully-glazed cladding solution, achieves an elegant and visually symmetrical intervention in the streetscape. The glazed section is, in fact, symmetrical, with adjustments for the variations in geometry reconciled at the end collars.

The essence of the bridge is 'transparency'. Coupled with the need to hide the differing floor levels on each side, this was what led to the design of a regular geometric structure with a tapered and sloping deck within. The hyperbolic paraboloidal shape was achieved by using a prestressed arrangement of circular, hollow-section struts and solid rod-ties between two circular lattice trusses.

The nine straight 114-millimetre (4^1/$_4$-inch) diameter tubes arranged around a horizontal axis give a sleeve of constantly varying diameter. Alternating with the tubes are an equal number of 28-millimetre

(1-inch) diameter rods, pretensioned so that no combination of loading can result in stress reversal and slackening of the rods. Analysis checks focused on the structure's sensitivity to frame buckling and parameters were adjusted to give the optimum solution. At both ends of the central section of the bridge, asymmetrically braced collars transfer loads to the supporting buildings. These collars also act as anchorage for the tension rods. The collars were site-bolted to the central section to facilitate erection and avoid costly in-situ welding. Due to programme constraints the collars had to be erected several months in advance of the middle section.

The structure of the deck of the internal walkway consists of steel joists spanning longitudinally onto slender steel crossbeams coincident with hoop frames at alternate node points on the primary structure. The level of this deck varies by some 1,200 millimetres (4 feet) along the length of the bridge.

Manchester is renowned throughout the UK, not entirely unjustly, as a place where it frequently rains. The architectural intent therefore was to protect shoppers from the weather by providing a fully enclosed bridge. This was achieved by cladding the end collars with decorative aluminium grilles enclosing weathertight inner membranes which in turn clad the openings used for natural ventilation. The solution for the glazed section was a series of triangular glass panels in ten different sizes which were laminated, heat-treated and positioned to create the three-dimensional curvature. The panels are fixed by clamping the corners of each one to the steelwork by means of cast stainless steel nodes. The issue of safety for maintenance staff on the glass was

Below: The deck is inclined, to match the floor levels of the shops on either side, but the envelope is on a level axis. The new shops replaced those destroyed by an IRA bomb in 1996.

Below: The Bridge of Sighs in Venice, built in 1600, shows an earlier architectural treatment of a bridge between two buildings.

Above: The bridge is highly transparent, which emphasizes the geometric shapes of the structure.

addressed by using polyester PET (polyethylene terephalate) interlayers. These extend beyond the corners of each glass panel in the form of rectangular tabs, which are mechanically restrained within the stainless steel nodes. Even if both layers of glass are broken, the interlayer is sufficiently strong to support the weight of a person safely for an extended period of time. To confirm the viability of this proposal, a rigorous testing regime was specified and implemented by the contractor.

The structure is exposed and transparent, and with its unique hourglass shape has become an icon within the city-centre regeneration project.

Conclusion

Manchester suffered severe trauma when an IRA bomb blew the heart out of the city centre. But the tragedy also offered an opportunity to revitalize the area and replace many of its uninspiring and dated structures with interesting and distinctive buildings. This footbridge, crossing a main pedestrian street and framing the views, is in some ways reminiscent of the famous Bridge of Sighs in Venice, but the Manchester structure represents a very different emotion – hope rather than despair.

> Hulme Arch Bridge

Manchester, UK, 1995–1997

In 1992 Hulme City Challenge was launched with the aim of regenerating this area of southern Manchester, which had become associated with poor housing and limited opportunities. Built in the 1960s on the site of Victorian terraces, Hulme's high-rise blocks were typical of public housing architecture at the time. The plan provided for replacement housing on a more traditional street pattern to help recreate pride in the community. Stretford Road, which had formerly linked the area to the city centre, had been bisected by a dual carriageway, Princess Road. This severance is thought to have contributed to the isolation and dereliction of Hulme. In reconnecting the route, Hulme Regeneration Ltd, the agency responsible for the regeneration plan, wished to create a landmark bridge that would act as a centrepiece and a visual marker of the area's revitalization.

A two-stage open design competition was held, and the six entrants chosen for the second stage were invited to develop their schemes further. The winning team of Chris Wilkinson Architects and Arup were selected in June 1995.

Opposite: The arch is a highly visible feature. Each half supports the bridge deck from one side.

Design

The design was both simple and unique. The bridge is supported by cables hung from a single, diagonal parabolic arch which appears as a gateway to motorists crossing the bridge on Stretford Road and those using the Princess Road dual carriageway below. It rises 25 metres (82 feet) above the bridge deck and was conceived as a bright, smooth, metallic structure. This effect was achieved by using a plated-steel box structure coated in aluminium paint.

The arch is trapezoidal throughout and varies from 3 metres (10 feet) wide and 0.7 metres (2 feet 3 inches) deep at the crown to 1.6 metres (5 feet 3 inches) wide and 1.5 metres (4 feet 11 inches) deep at its springings. This variation, combined with the essential asymmetry of the bridge as a whole, results in a live structure which is perceived as different from each new viewpoint. For observers passing through the arch there is thus the added interest of a constantly changing form.

The arch is supported on substantial ground-bearing foundation blocks each made with approximately 300 cubic metres (400 cubic yards) of concrete, anchored at each end with 32 high-tensile stainless steel bars. The shape of the arch follows the thrust line generated by the in-plane effects of the permanent load from the cables reasonably well. However, due to the arrangement of the cables, the arch is subjected to very considerable out-of-plane asymmetric bending effects. These effects governed the design, in some cases using up to 75 per cent of the section capacity, and as a result the arch functions more as a laterally loaded bending member than a conventional arch.

The arch section was designed to be fabricated in two pieces, like a coffin. The soffit and two side plates, together with the cable lugs and associated internal diaphragms, form the box, and the independently stiffened top plate becomes the lid, connected to the box with an external, longitudinal seam weld on each side.

The 52-metre (170-foot) span bridge deck is hung from the arch by 22 spiral-strand diagonal cables, 51 millimetres (2 inches) in diameter, each with a minimum breaking load of 216 tonnes (213 tons). The cables are asymmetrically arranged and fan out in opposing directions, with each side of the deck connected to a different half of the arch. The resulting arrangement encloses the space above the bridge and contributes to the overall appearance of the structure. The bridge deck is a composite concete slab laid over 17 transverse girders spanning on either side to two steel edge-beams, connected to the stay cables via outrigger brackets. The edge of the structure is given visual continuity by a tubular-steel nosing supported outside the cable brackets.

Bollards along the inside of the footways on either side of the deck prevent high-sided vehicles riding up onto the footways and snagging on the inclined cables. The arch is lit by uplighters set in the side slopes of the Princess Road cutting, and red light-emitting diodes set into the bollards illuminate the footways, providing a contrast to the silver lighting of the arch.

Opposite: The box profile of the arch gives clean details and conveys simplicity.

Above: The two prefabricated halves of the arch were lifted and joined in a short closure of the main road.

Below: The arch segments were placed in the central reservation of Princess Road while awaiting erection.

Construction

Princess Road is an important route from the city centre to Manchester Airport, and therefore possessions for construction of the bridge were limited. Use was made of the wide central reservation of Princess Road to assemble off-site fabrications into larger elements for lifting into place. The deck and the arch were each installed in a single weekend possession. The deck was prefabricated as single beams and then assembled into three sections – each 17 metres by 17 metres (56 feet by 56 feet) – in the central reservation. These were then craned into position, resting on the abutments and four temporary trestles in the central reservation. Permanent formwork for casting the deck concrete was placed in position during the same possession.

The arch was fabricated from six pieces approximately 15 metres (49 feet) long, which were welded together in the central reservation to form two 80-tonne (79-ton) halves. These were erected in a tandem lift with two 500-tonne (492-ton) cranes. The cables were rigged and stressed in a third weekend possession of Princess Road.

The bridge was formally opened on 10 May 1997 by Sir Alex Ferguson (manager of Manchester United Football Club), in the presence of Sir Bobby Charlton (a member of the England football side that won the 1966 World Cup) and civic dignitaries. The first vehicle to cross the bridge was the first Rolls Royce ever made, which had been manufactured nearby. The opening of the bridge marked the five-year point in the regeneration of the Hulme area and was celebrated by a spectacular evening firework display at the site.

Conclusion

The requirements for this bridge were symbolic as well as practical. The redevelopment of Hulme in the 1960s had created a district of high-rise buildings which, although providing modern plumbing, destroyed the community. In practical terms, regeneration of the area required the demolition of the high-rise blocks as well as much of the more recent building stock. The creation of the landmark bridge, using a design unique at the time, not only provided the physical connection between Hulme and the city of Manchester, but also engendered pride in the area. This shows that the design of transport facilities is not just about the physical structures, but also about the part that engineering and architecture can play in the creation of a community.

Right: The arrangement of the cables gives different views from different angles.

> Chapter 4
**Transport Planning
and Special Projects**

> Transport for the 2000 Olympics

Sydney, Australia, 1996–2000

The 2000 Olympic Games were held in Sydney from 15 September to 1 October, with the Paralympic Games following from 18 to 29 October. The transport requirements for these events included the planning of travel facilities for temporary venues and the logistics and security aspects associated with ensuring that spectators, competitors and officials could arrive at and leave events on time. There was also considerable investment in permanent sports facilities and transport infrastructure – including the new Olympic Park railway station and the expansion of Sydney Airport – and these form part of the legacy of the Games.

The demands of the Olympic Games put pressure on all aspects of the city's transport and management systems. During the bid to host the games, Sydney had promised a 'green' Olympics; the city adopted policies to combat the growing imbalance between its road and public transport systems so as to cut pollution and protect the environment. The overall strategy was therefore built around access to the various venues by public transport. Some venues normally accessible by car became 'car free' because only public transport was able to cope with the numbers involved. In addition, it was vital that roads were kept free of congestion for security and public-safety reasons and to provide easy access for the emergency services. Since the massacre at the Munich Olympics in 1972, the security of competitors and team officials has been of paramount concern and they are now required to travel to competition venues separately from the general public. At the Sydney Olympics they were transported by bus.

Previous page: Picture shows the roof structure of the CargoLifter Airship Hangar.

Opposite: At the Olympic venues it was necessary to make arrangements to handle large numbers of visitors in a short time.

Planning

The policy document 'Action for Transport 2010', adopted by the New South Wales Government in 1998, contained a proposal to fund new rail projects as well as urban light rail, rapid bus-only transit ways and cross-regional bus services. The annual budget was A$300 million (US$181 million/£115 million). Many of these projects, including the Olympic Park railway station and the Airport Rail Link, were to be completed in time for the Olympics.

Transport planning involved the movement of not just vehicles but also pedestrians. In addition to the workforce and the spectators, there were 10,500 athletes, 5,100 officials, 16,200 media personnel and 50,000 volunteer Olympic Hosts for 300 events to be considered. All venues had to be designed for the safe exit of these people in the event of a fire or emergency. And with the exceptional demands of the Olympics, there was a need to check the capacity of emergency exits at existing as well as new and temporary venues to ensure public safety. Railway stations would also be subject to unusual demands and so safe working practices had to be established at these locations, too.

Organization

The events were run by the Sydney Organizing Committee for the Olympic Games (SOCOG) and the Sydney Paralympics Organizing Committee (SPOC), at venues planned and constructed by the Olympic Co-ordination Authority (OCA). Transport was provided by the Olympic Roads and Transport Authority (ORTA), who established a Transport Management Centre (TMC) to plan and manage the system.

The bus fleet comprised 3,350 dedicated buses with 4,500 drivers. A holiday was declared for schools during the events, allowing school buses to join the Olympic fleet. For the Paralympics, all available low-floor buses were taken from normal services to provide transport for the competitors. An Olympic Bus Depot was established at Regent's Park, with capacity for over 1,000 buses to be parked, washed, security checked, refuelled and repaired on the site at any one time, 24 hours a day.

Usage of the CityRail network rose from the normal 14 millon to 34 million trips over the 17-day period of the Games. Of these trips, 31 per cent involved travel to Olympic venues, 34 per cent Olympic sightseeing, and 35 per cent non-Olympic trips. The Olympic Park railway station was planned to cope with 6.8 million trips, an average of 400,000 per day on 419 timetabled trains.

On the peak day of the Olympics 400,345 spectators visited Homebush Park and an estimated 150,000 went to other venues. In addition, events such as the Marathon attracted huge unticketed crowds, and large numbers of people congregated at remote outdoor venues to watch the Olympics on large screens. An estimated 400,000 additional people visited the city centre every day of the Games. The number of Transport Authority staff grew to 19,403 – including bus drivers and volunteers – to cope with the exceptional demand.

Opposite: Several venues were near to each other in the Olympic Park, and management arrangements had to ensure that crowd movements were separated.

Right: Car access was prohibited, and spectators arrived by train or bus or on foot.

Three-quarters of the population of Australia live more than six hours' drive from Sydney. The city's only international airport was therefore also the major point of arrival for domestic spectators. In total it was estimated that 240,000 international and 180,000 domestic travellers (competitors, officials, press and spectators) arrived and departed from Sydney Airport. Peak passenger movements increased from 66,000 to 150,000 per day. Additional demands requiring special attention included 260 horses for the equestrian events and 4,000 disabled athletes for the Paralympics. Aircraft movements increased to 1,100 per day, close to the number at some of the largest airports in the world.

Various airport improvements were brought forward and completed before the Olympics, including additional apron space and contact gates. The airport has been connected to the city by underground rail since May 2000, and this helped to ease the landside transportation problems.

The Sydney Olympic Park provided new venues for the opening and closing ceremonies, and for archery, athletics, badminton, baseball, basketball, diving, football, gymnastics, handball, hockey, modern pentathlon, swimming, table tennis, tae kwon do, tennis, volleyball and water polo. There were seven other main venues in the Sydney

region. All public parking was withdrawn from the venue sites and ticket holders were granted free travel on the Olympic transport system, comprising over 300 CityRail stations, Homebush Bay regional bus routes and shuttle buses connecting CityRail stations to the Olympic venues.

The Olympic Park railway station

The new Olympic Park station is the gateway to the Homebush Bay Redevelopment. It was an integral part of the promises made to the International Olympics Committee at bid stage and forms part of the Metropolitan network. For the Olympic Games it was predicted that the station would need to cater for 50,000 passengers per hour, a very considerable throughput for a station with only two platforms. There are platforms on both sides of both tracks so that all arriving passengers alight at the central island platform and departing passengers wait on the two outer platforms. The operational plan required double-decker trains with a crush-load capacity of 1,700 passengers to arrive and depart at two-minute intervals. Such flows are rarely achieved at major multi-platform stations in large cities, and are similar to the flows through the main Sydney Town Hall station in the Central Business District during the 3½-hour morning peak.

Above: Special arrangements were made for safe access to Bondi Beach during construction of the beach volleyball stadium.

Opposite: Trains had to be carefully scheduled at the Olympic Park station to ensure that platforms did not become dangerously overloaded.

The station layout was refined using pedestrian modelling software which analyses flow rates, crowd densities and platform congestion. To calibrate the software, loading and unloading trials were undertaken with the type of double-decker trains to be adopted, using volunteer members of the public, including all age groups and wheelchair users. Observations were also made at other busy stations in the city to ensure that the model parameters reflected the behaviour of crowds in Sydney. Crowd behaviour varies from country to country and city to city as a reflection of different cultural norms, and so models always have to be recalibrated for local conditions.

In order to minimize delays to trains and allow the maximum number of trains to use the station at peak times, it was important to minimize the dwell time. When a train arrives, the exit doors onto the central platform are opened. Once most passengers have alighted, the entrance doors to the outer platform are opened. As soon as they are satisfied that all passengers have left the train, they close the doors onto the central platform. This system reduces conflict between arriving and departing passengers and allows the train to fill up more easily to crush-loading point. Access to the underground departure platforms is controlled at the ticket barriers at ground level. If too many passengers were to be allowed onto the platform there is an increased risk of someone being pushed onto the tracks. The departure of the train would also be delayed as passengers seek to

force their way onto trains that are already full. A predetermined crowd-management strategy is therefore implemented by station staff to ensure both safe operation and maximum throughput.

The central island platform was designed to cope with 900 exiting passengers per minute (54,000 per hour). To achieve this, it has four escalators, two lifts and four sets of 3.5-metre (11½-foot) wide stairs. The capacity of the exits was also checked against standards set for fire- and life-safety issues to ensure that appropriate safety criteria were met. Few major-city stations in the world are required to deal with these design flows, and most of those have many platforms. The maximum 27,000 passengers per hour in each direction is the capacity limit of most urban metro systems, but these rarely have to accommodate all passengers boarding or alighting at a single station.

Conclusion

The world inevitably focused on the Games' athletic feats without paying much attention to the transportation 'hardware' they required – railway stations, bridges and buses. However, the enduring 'green' legacy for Sydney and international visitors is the city's new emphasis on public transport, making this (and walking) an enjoyable way to get to cultural and sporting events within the metropolitan area. The shift from private to public transport – started during the two years of Olympic test events – has already exceeded all expectations. A major factor in this shift has been public acceptance of the new transport systems and management.

The Sydney Olympic Games were recognized as the most successful ever. They engendered tremendous interest and raised the profile of both Sydney and Australia throughout the world. After the transport problems of the Atlanta Games in 1996, Sydney was determined to show the world that Australians know how to organize a party. They certainly did that, and the transport arrangements played an essential role in its success.

> International Border Crossings and Transport

Central and Eastern Europe, ongoing since 1993

After the collapse of the communist regimes in Eastern Europe, major changes occurred in the patterns of movement of goods due to the fact that trade was no longer concentrated on Russia, and traffic with the European Union (EU) was increasing dramatically. Populations in the newly democratic countries had high expectations that their standards of living would rise rapidly to those in the West, but neither the physical nor the social infrastructure existed to create the necessary wealth. The EU responded with programmes designed to assist these countries with the reconstruction of their economies. Its main concerns were trading relations and the creation of a common market for goods and services in Europe. The two programmes described here illustrate the nature and difficulties of creating the conditions for free movement of goods.

The first project relates to the problems created at national borders when major political and structural change takes place. The second covers the assistance given to railway administrations to ensure that long-distance freight trains can compete with road haulage.

Opposite: Jagodin Bridge is on the Ukraine-Poland border.

Border crossings

In 1993 Arup undertook an appraisal mission for the EU to evaluate the international transport needs of the former Warsaw Pact countries of Central Europe. These countries benefit from the EU's PHARE programme, which gives support in the restructuring of their economies and the modernization of their social and physical infrastructure. The aim is to secure sufficiently stable and strong economies in preparation for EU entry. The beneficiaries at the time of Arup's mission were Albania, Bulgaria, Romania, Slovenia, Hungary, Slovakia, the Czech Republic, Poland, Lithuania, Latvia and Estonia. Most of the former Yugoslavia was excluded due to the civil war, but Bosnia Herzegovinia and Croatia have since been added to the programme.

In all countries there were serious transport problems caused by decaying infrastructure and poor management. To help rectify this situation the PHARE programme provides technical assistance to support the national administrations and to evaluate projects for funding from the European Bank for Reconstruction and Development, the European Investment Bank and private funding sources. The Regional Transport Programme was concerned with cross-border transport.

The mission found that a major problem was the cross-border infrastructure, both physical and administrative. Many border crossings had been closed during communist rule and there were few facilities for international trade, such as customs posts. Where they did exist they more often than not had inadequate facilities for inspecting road and rail vehicles to enable border taxes to be levied and to detect the smuggling of people and goods. Where borders crossed rivers, most bridges and ferries had fallen into disrepair.

The patterns of trade changed as links with the former Soviet Union weakened and trade with Western Europe rapidly increased. At the same time, cross-border trade between Central European states also increased, particularly road traffic. Delays at border crossings were high, sometimes lasting several days, and queues of 10 kilometres

(6 miles) were commonplace. One even stretched 30 kilometres (19 miles). The economic loss and environmental degradation caused by the congestion at borders were adversely affecting progress towards more liberal trading relationships between the states themselves and with the countries of the EU.

The main problems identified at borders were:

– inadequate infrastructure
– outdated customs-clearance controls and procedures
– poor administration and traffic management
– lack of co-ordination between authorities at borders
– insufficiently motivated staff

Arup was appointed initially for a study of border infrastructure problems in the Balkans region. Because routes through the former Yugoslavia were closed due to the conflict, high volumes of traffic were seeking routes through Hungary, Romania and Bulgaria to Turkey and the Black Sea. Other traffic was passing through Italy and crossing from Bari to Durres in Albania. Trucks were then routing through Skopje in Macedonia and Sofia in Bulgaria. In Albania, where border crossings had been closed for 50 years, there were no facilities at newly opened crossing points and the approach roads were in very poor condition. The first task of the study was to identify short-term measures to reduce bottlenecks. Arup's work was later extended to cover all of Central Europe, and all the PHARE countries became involved in the programme.

The normal procedure at border posts is to check a driver's passport and visa first and then inspect the vehicle. This may involve weighing it to assess duties, and bonds or other documents may need to be completed before the vehicle can proceed. Where waiting areas are inadequate, vehicles which could be checked quickly are often delayed by vehicles requiring a more thorough check. Frequently, this procedure is then duplicated on the other side of the border, meaning that a vehicle crossing several countries may be checked many times.

Although international customs clearance procedures should reduce the amount of duplication, vehicles with the correct clearance can still be held in queues because there is no space for them to pass other vehicles.

At many of the borders a large proportion of the vehicles are engaged in the 'suitcase trade', with small-scale entrepreneurs carrying goods in cars and small vans. The effort involved in inspecting them is disproportionate to the revenue obtained from them in customs duties.

Although inadequate infrastructure was generally held to be the core reason for congestion at border posts, this was probably never the case. Once the worst shortcomings of the infrastructure had been addressed, attention shifted to its administration and organization. A legacy of poor management practices existed, and controls were operated differently in the various countries. Excessive time was sometimes taken for routine clearances; petty corruption was rife at some border crossings, and some were alleged to be controlled by criminal elements. These circumstances gave little incentive to reduce queues because delayed drivers were more likely to be willing to pay bribes in order to secure rapid clearance. Staff from the British Customs service augmented the Arup team to advise on international customs-clearance matters and other institutional issues.

In many cases, noise and air pollution from lorries queuing at borders close to habitation were creating environmental problems. Nuisance was also caused where there were inadequate toilet and washing facilities for waiting drivers. However, some local economies were boosted by trade with the drivers in the queues and in these instances there was resistance to improved border facilities. There are other potential employment consequences. For example, at the border of the former Soviet Union where the railway track gauge changes, large numbers of people are employed in changing axles on trains. The introduction of mechanized axle-changing equipment would result in significant redundancies.

In some cases new borders had to be marked. The Baltic states of Lithuania, Latvia and Estonia were incorporated into the USSR in 1940. Borders between Russia and the Baltic states, and between the states themselves, had therefore not existed for over half a century and were no longer physically marked.

The border-crossings programme was directly funded by the EU, and over the period 1996–2000 total investment, including co-funding from the governments concerned, was around €350 million. Local firms were employed to design and supervise works under the management of Arup, and 109 contracts were let under the FIDIC (*Fédération*

Below: Leushen, Moldova. The entire land border of the state of Moldova adjoins Romania.

Below bottom: Narva, Estonia. The Baltic States had been absorbed into the Soviet Union. New border facilities were therefore needed when they became independent.

Below: Domoshevo, Belarus. The Poland border looking towards Belarus.

Below bottom: Kamenny Log, Belarus. The border crossing to Medininkai, in Lithuania.

Opposite: Kuznica, Poland. When Poland joins the European Union in 2004, its eastern border will become the boundary of the EU.

Internationale des Ingénieurs-Conseils) Conditions of Contract to meet EU procurement rules. The economic benefits of the programme were also assessed. For example, it was calculated that if delays could be reduced to an average of two hours at one particular crossing – Swiecko, on the Poland/Germany border – the benefit would be €320,000 (US$349,000/£221,700) per day.

When Central European countries join the EU, their borders within the EU will no longer exist since the Schengen Agreement provides for a passport- and customs-free union. The eastern boundaries of those states will therefore become the boundaries of the EU, and immigration and trade controls will need to be managed at those borders. With the great disparity of wealth between the countries of the EU and those of the CIS (the Commonwealth of Independent States, derived from the former Soviet Union) such as Belarus, Ukraine, Moldova and Russia, strict controls on the smuggling of goods and people will be required. In 1999 a €50 million (US$54 million/£34.5 million) programme to improve security along these borders was commenced.

The overall programme stretched from the Black Sea to the Arctic Circle. Most of the projects involved new or improved facilities at border posts – buildings, weighbridges, parking areas and so on. Others improved road access to the borders, and a new bridge was constructed over the River Bug at Jagodin on the border between Poland and Ukraine.

Trans European Rail Freight Freeways

Major changes have taken place in patterns of trade and the organization of their transport systems in the Central and Eastern European Countries (CEECs) as a result of the radical political upheaval that followed the removal of the Berlin wall and the downfall of communist regimes. State direction had favoured railway transport for both passengers and freight, but the introduction of free markets brought about significant changes to both the nature and economics of freight movements. In particular, the increase in flexibility and speed offered by road transport resulted in a reduction in rail traffic.

These changes reflected the situation in Western Europe, where between 1970 and 1994 rail freight lost half its market share, mainly to road haulage. Rail is a more efficient carrier than road for long hauls, but where an international border crossing is involved, problems can arise. Delays are caused by customs and immigration procedures and by changes of train operators and infrastructure managers. The marketing of freight services can also be poorly co-ordinated.

Customs authorities often require freight trains running through international borders to be checked several times. A change in rail administrations will often require a change of locomotive and drivers, as well as bureaucratic procedures to record the wagons moving from one country to another. The resulting delays increase both journey times for goods and the risk of loss. With road transport there are fewer problems with security and customs – a major concern to consignees – because the driver is responsible for the load.

The reasons for the decline in rail freight relate mostly to the various aspects of 'quality of service'. For international transits involving more than one railway administration, the problems referred to above make these aspects significantly worse. The European Conference of Ministers of Transport identified the main problems as:

– technical incompatibilities
– differences in organization between national railways
– financial constraints and issues at national level
– balance of payments considerations
– poor inter-railway communication and load tracking

A train from Rotterdam to Budapest, for example, has to cross at least three national borders using track electrified to three or four different systems, with checks and paperwork at most borders. Delays can occur when locomotives are changed or awaiting train paths between passenger and local freight trains on each national system. In addition, the weight of the train is limited to that which the part of the route with the lowest carrying capacity can accommodate. The limitation may apply to the maximum axle load permitted on the track or to the weight of the whole train, and is determined by the gradients on the line and the power of the available locomotive. The contract between the railways and the shipper will be made by one railway on behalf of others, and each railway administration may have different commercial priorities.

The strategy of the EU is to maximize rail-freight usage – for environmental reasons, because rail is more energy efficient and less polluting than road haulage; for economic reasons, because well-managed railways can provide greater efficiency, particularly for long-distance freight.

In Western Europe the EU has been promoting the concept of a Trans European Rail Freight Freeway (TERFF), which provides simplified procedures and operations for companies wishing to send trains through several countries. The first – between Belgium, France, Italy and Spain – was established in 1997. The second tranche of routes includes links between Scandinavia and Germany, Italy and Austria, and the UK and Hungary. The concept's main concern is institutional framework rather than infrastructure, reflecting the nature of many of the problems. The basic requirements of the TERFF concept are:

– a voluntary agreement between infrastructure managers (IMs) along existing international routes to operate a freeway
– a set of attractive train paths available to purchase at short notice from a 'one stop shop' (OSS)
– IM agreements to pass control of TERFF train paths to the OSS to allow them to be marketed at a throughout price under the terms of a single contract
– management by the OSS, including monitoring the progress of trains and dealing with delays
– customers who are a Licensed Railway Undertaking (LRU) providing international rail-freight operations
– by TERFF customers to the OSS for the use of TERFF train paths, and OSS payment to the IMs

Arup's study of the extension to TERFF into CEECs identified several corridors. North-south corridors were considered from the Polish Baltic coast to the Czech Republic and Slovakia. Two east-west groups of routes were studied: one from Germany to Belarus through Poland, and one from Austria and Slovenia to the Black Sea and Turkey through Hungary, Romania and Bulgaria.

The CEEC railway track and signalling were generally consistent with Western European technical standards, but the wagon fleet was in poor condition, resulting in problems of acceptance by EU railways. The CEECs concerned operate on European-gauge tracks, with a transfer to the wider Russian gauge at the former Soviet Union border. Most countries had already established open access for freight-train operators, where other train operators can apply to run their own trains over the tracks of the national railway. New open-access operators were successfully taking market share from the national railways, which were using freight operations to subsidize passenger services.

Following the study, a conference for delegates from governments and railway administrations was held at which there was considerable interest in establishing TERFFs in CEECs. To achieve this it would be necessary to:

– create a mechanism for financing and procuring modern wagons
– address border crossing delays
– improve timetable co-ordination
– develop a system for tracking and tracing wagons and loads to meet the needs of customers

Conclusion

These projects are a clear demonstration that in order to achieve good transport facilities it is necessary to create efficient management and administration systems as well as the infrastructure. The examples described here show how Western countries are helping to implement the reforms that are necessary in Central Europe.

Above: Eastern European railway networks had to reform their operations and cross-border co-operation arrangements to compete with road transport in a competitive market.

> Traffic Calming Scheme

Brooklyn, New York, USA, 1999–2000

In most cities an increase in traffic is causing congestion, safety and pollution problems in downtown areas. In Brooklyn these problems are accentuated by four particular characteristics of the area. Firstly, there is a shortage of parking spaces since most residential properties have no off-street parking. Secondly, there are many subway stations in Brooklyn and commuters from Long Island come into the area to take the subway to Manhattan rather than queue for the bridges or tunnels. Thirdly, the regeneration of the downtown Brooklyn area – developed as a partnership between the education, business and government sectors – has been very successful, increasing employment by 25,000. Finally, the location, at the eastern end of the Manhattan and Brooklyn Bridges, causes traffic headed into Manhattan to concentrate in the downtown Brooklyn area.

The Gowanus and Brooklyn-Queens Expressways were designed to remove through traffic from the area, but because these routes are now congested drivers are seeking faster routes through local streets. The community in Brooklyn is mixed, both ethnically and economically, but all parts of it have been affected by the decline in local living conditions resulting from problems caused by excess traffic.

Opposite: Kerb build-outs and road humps at intersections slow traffic and shorten pedestrian crossing distances.

Inception of the project

After studying concepts of 'traffic calming' developed in Europe and Australasia (but virtually unknown in the USA), local politicians and residents groups decided to develop a project to apply these techniques in Brooklyn.

'Community involvement' was at the core of the Downtown Brooklyn Traffic Calming Project, and community representatives appointed by the borough president and local councilman were members of the panel that chose Arup for the project. The project team comprised consultant advisors from the UK and Australia, as well as local traffic engineers, a landscape architect, an urban designer and a public outreach consultant. The international experience of the team was important for this project, which was seen by the community and the New York City administration as signalling a new direction in traffic management. Parallel initiatives were taking place in Manhattan to improve local pedestrian conditions in Times Square and sections of Broadway, but the Brooklyn project was the only one to cover a wide area including both residential districts and downtown.

A community-based Task Force chaired by the Brooklyn Borough President monitored the study. A Technical Advisory Committee was chaired by the New York City Department of Transportation (NYCDOT), which consisted of interested city agencies, including the police and fire departments.

Traffic calming and the objectives of the project

Traffic calming is the term used to describe actions to reduce the intrusion of motor traffic into urban life. The best-known manifestation of traffic calming is the road hump or 'sleeping policeman', which has been used in a few places in New York City, mainly outside schools. There is, however, a wide range of physical, management and educational measures within the traffic calming toolkit.

The scope of the project was to prepare a series of measures to be implemented in the area as a pilot programme, and then to assess both their technical suitability and the response from the local community and prepare a masterplan for amelioration of traffic conditions throughout the area. The objectives were to:

– improve pedestrian safety and access (including safer crossings at problem locations), reduce vehicular speeds and enhance mobility between neighbourhoods
– reduce unwanted traffic impacts, including congestion, inappropriate vehicle volumes, speeding, noise, air pollution and damage to infrastructure
– preserve and improve civic, commercial and residential area access by providing a traffic-calmed street network for improved connectivity between these destinations
– protect the unique character of historic residential communities

A complementary list of objectives flowed from the outreach process undertaken for the project. These included the desire to:

– improve pedestrian circulation and safety
– develop the local cycling network
– manage truck access and routing whilst reducing the impact of trucks on the community
– maintain local traffic permeability
– maintain or enhance emergency-vehicle access

Opposite: Speeding cut-through traffic in residential areas threatens the safety of residents.

Below: In busy urban streets active traffic management is needed if people and cars are to co-exist comfortably.

Whilst area-wide traffic calming is relatively new to North America, particularly on the east coast, the techniques have been developed over the last 40 years in Europe and in Australia, where the urban form is similar to that of US cities. The adoption of the philosophy that road traffic should be subservient to a city's design objectives rather than the source of them, leads to a new urban architecture where the spaces between buildings are as important as the buildings themselves – and, frequently, more important. The vision is often driven by community pressures, and city officials more accustomed to pressure to accommodate car use are now being urged to adopt a strategy to mitigate its effects.

As the project developed the management framework altered. The community came to the project thinking it was about forcing NYCDOT to install speed humps in their pet locations, and NYCDOT were initially hostile to this approach. However, using the experience of the team, the project was refocused on the creation of the framework for a co-ordinated plan that could be developed and implemented.

An important part of the management approach was to define the role of streets. A three-part classification appropriate to Brooklyn was developed, comprising Traffic Streets, being the main arteries, Community Streets, where local shops and community facilities were located, and Living Streets, with no through-traffic function and normally accessed through Community Streets. This functional distinction allowed differing objectives to be set for each category of streets, and as a result it became possible to close the gap in thinking between the community and the city traffic engineers. This was important, for the appropriate traffic-engineering response depended on agreed objectives for the street.

Opposite: Cyclists and pedestrians are vulnerable road users. Traffic calming aims to improve their safety.

Above: Parking close to intersections masks pedestrians trying to cross the road.

Above centre: Parking and driving in cycle lanes reduces their safety benefits for cyclists.

Above right: Wide crosswalks are frightening for pedestrians. Often the roadway can be narrowed without affecting traffic capacity.

The pilot project

A variety of measures were evaluated for use in Brooklyn, including physical management and education measures.

Many of the physical measures concerned the reclamation of road space for pedestrians, allowing for tree planting, seating and sculpture as well as increasing pedestrian security. For example, kerb neck-downs at intersections where there is a parking lane on the approach allow pedestrians to remain on the kerb while waiting to cross. Pedestrians, particularly in New York, tend to stand in the carriageway in order to see oncoming traffic, and so the neck-down reduces the risk of them being struck by turning vehicles.

In some locations it was found that pedestrian crossing distances could be substantially reduced. For example, on Tillary Street there are five lanes approaching the junction with Adams Street: two for left turns, one for right turns and two for traffic going straight on. Downstream of the junction only two lanes are required – to match the two approaching straight-on lanes – and the width here could be reduced by up to 60 per cent to make pedestrian crossing easier and safer. Following the pilot programme it is expected changes of this nature will be implemented, together with landscaping of the reclaimed space to soften the harsh impact of the urban environment at such locations.

Management measures can be equally simple. On DeKalb Avenue, signals were retimed so that drivers observing the speed limit of 40 kilometres per hour (25 miles per hour) would have a continuous 'Green Wave', whereas drivers exceeding the speed limit would always face a red light at the next set of signals. Unfortunately the pilot showed that this had no beneficial effect on traffic speeds – perhaps because there was no indication that the 'Green Wave' existed. A more successful manipulation of traffic signals was the 'Leading Time Interval'. This allowed pedestrians to start to cross a wide road before the traffic received a green light, thus forcing turning traffic to give way to pedestrians already on the crossing. This was highly popular and was perceived to give a major safety benefit, although at the expense of some traffic capacity. At another junction an 'All Pedestrian' phase was introduced into the signal timings. This involved all pedestrians

receiving the 'Walk' signal for a period when all vehicles were held by a red light. This was not well understood by pedestrians or drivers and probably needs extra signing to be effective.

In another scheme, blue surfacing was applied to a cycle lane to deter both moving and stationary cars encroaching onto it. The scheme was particularly successful and after a few months it was extended at the request of users.

Conclusion

This project, designed to improve safety and the environment over a large downtown and residential area, demonstrated that importing design solutions from another continent is not as simple as it might seem. Pedestrians and motorists become used to the informal as well as formal conventions of the traffic regime in their city and do not readily understand measures that may be commonplace elsewhere. The project was a valuable exercise in bringing together community activists and city transport officials to plan and execute an effective programme of action. The lessons imported covered not just traffic engineering techniques, but also the management of city-street networks with a structured set of local objectives. Pilot projects demonstrated the potential of this approach, and it will be interesting to see if New Yorkers come to terms with the behavioural changes necessary to achieve the quality of street environment assumed to be normal in many Western European countries.

> CargoLifter Airship Hangar

Brand, Brandenburg, Germany, 1997–2000

Airship development was active in the USA, Britain and Germany in the first half of the 20th century, but a combination of the disasters to the R101 and the Hindenburg and the rapidly advancing technology of heavier-than-air aircraft spelt the end for commercial airships. Since 1945 they have had a limited use in advertising and for military applications such as submarine hunting. Now airships are once again being developed in Germany. A company called CargoLifter AG will use them to lift and transport goods of up to 160 tonnes (157 tons) over long distances. It has built a hangar to produce and maintain the first two of a new generation of CL160 helium-filled airships or 'blimps' – so called because they have no rigid skeleton. No internal structure means less self-weight to lift, so they will be able to carry far more than the 60 or so tonnes (59 tons) of the largest pre-Second World War Zeppelins.

The hangar is in Brand, some 50 kilometres (30 miles) south of Berlin, and the project comprises the hangar itself, several other buildings for component production, and a visitors' centre. The masterplan was developed by architects SIAT Architektur + Technik of Munich, who co-ordinated the planning and were supported from an early stage by a design team including Arup as structural engineer. Though it is an industrial building, the architecture of the CargoLifter hangar is of considerable importance, both functionally and visually, due to its size and impact.

Opposite: A smaller version of the airships to be produced here is dwarfed by the scale of the hangar, which is designed to hold two massive 160-tonne-payload blimps.

Like railway stations, bridges, exhibition halls, towers and other examples of industrial buildings, airship hangars are outstanding and fascinating examples of structural engineering. However, to a greater degree than these types of structure, in the past hangars have been designed purely with economics and function in mind, often ignoring aesthetics.

In Germany, airship hangars were built only between 1898 and 1938 and there is no surviving example nor any personal recall of them, and documentation is very scarce. Nevertheless, the CargoLifter hangar continues the clear lineage of traditional German airship-hangar technology, incorporating developments that have taken place in the design, detailing and construction of large-span structures over the intervening 60 years. For example, research on hangars and structural membranes by Frei Otto and Berthold Burkhardt at the Institute for Lightweight Structures (IL) at Stuttgart University during the 1970s and 1980s was repeatedly referred to during the design. Biaxial tests at the IL – as part of the SFB64 research programme – led on from earlier testing for zeppelin skins and are the present-day basis for the structural use of membranes.

The CL160s, being blimps, will be rounder and more compact than zeppelins, so the hangar is correspondingly wider and higher but not much longer than the largest of the earlier hangars. In the past, various types of hangar doors were developed and built – sliding, swivelling and rotating, foldable and sliding – but those for the CargoLifter hangar are absolutely new in concept. It is worth noting that, at the same time, Arup was developing a retractable roof for the Miller Park Stadium in Milwaukee, USA, which is also segmented in single Z-shaped elements rotating around a single central point.

Planning

Based on planning requirements for the CL160, a building envelope was developed which minimizes the building volume and surface and accommodates two CL160s alongside each other. The building type relates to the airship theme in shape and choice of material, and utilizes construction techniques developed from those used on past airship hangars. The envelope guarantees an economic, functionally efficient and aesthetically pleasing building, in which risk factors are minimized within the available timescale.

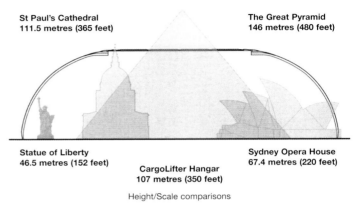

St Paul's Cathedral
111.5 metres (365 feet)

The Great Pyramid
146 metres (480 feet)

Statue of Liberty
46.5 metres (152 feet)

CargoLifter Hangar
107 metres (350 feet)

Sydney Opera House
67.4 metres (220 feet)

Height/Scale comparisons

Above: Comparisons with the Statue of Liberty, St Paul's Cathedral, The Great Pyramid and Sydney Opera House show the massive scale of the hangar.

Below: The profile of the hangar is defined by the full-height doors at the ends, each of which is a quarter of a sphere.

Opposite: Between the top chords, external props restrain any torsion in the arches induced by the eccentrically loaded membrane.

The building is 363 metres (1,190 feet) long and 225 metres (740 feet) wide, its most notable feature being the large semicircles at each end. Their 100-metre (330-foot) radius makes the central part of the building 'only' 160 metres (525 feet) long, which in section also forms a semicircle with a height of 107 metres (350 feet). This shape will snugly enclose the airships, but with ample tolerance. The segmented 'clamshell' doors at both ends of the building, each with two fixed and six moving elements, enclose the larger part of the hall. The additional cost of the doors' structure was offset by savings achieved on the smaller central part of the hangar and a reduction in floor-slab area – both of which resulted from the chosen door structure. In addition, the aerodynamic outline, closely matching the clearance shape for two airships, minimized wind load on the structure. The hangar covers an area of 66,000 square metres (710,000 square feet) – completely free of internal support – and encloses a volume of around 5 million cubic metres (180 million cubic feet).

The cylindrical central part divides into four bays covered with a translucent fabric and supported by five steel arches at 35-metre (115-foot) centres, springing off concrete plinths which also act as covered entrances. These arches have a clear glass roof between their top chords to allow daylight into the building, and through these 'arches of light' people can be directed straight to the nearest exit in the event of an emergency.

Two-storey-high concrete 'buttresses' along both sides of the production floor house facilities for 250 employees, laboratories, and offices for a further 75. The hangar has underfloor heating, plus radiant panels hanging from the steel arches in the side areas.

Early on, the IFI institute from Aachen carried out wind-tunnel tests for the membrane and door structures. In general, all the concept ideas, solutions, design methods and analyses developed by the design team were confirmed by contractors and the checking engineer (Prüfingenieur), as well as by the wind tunnel test carried out very early on in the process.

Structural concept and loading assumptions

Burkhardt and Angelika Osswald surveyed and identified various structural systems used in airship hangar design. From this it became clear that a fully moment-stiff arch structure most closely resembles the early reinforced-concrete halls in shape, size and structural system. A semicircle without hinges was a new structural shape. Structures with hinges were used in the past because they were simpler to design and easier to build. Now the CargoLifter has reverted to a rigid hingeless arch structure because it requires less steel.

The arches of the building spring from concrete plinths, which also act as covered entrances to protect against avalanches of snow from the 107-metre (350-foot) high fabric roof. These bases are founded on large concrete pad footings designed to limit settlement to 30 millimetres (1¼ inch), and to avoid sliding due to horizontal wind loads and thrusts from the arches. There are no additional tension members tying the bases. The steel arches have internal cross-bracing between them and props on the outside to avoid overall torsional buckling. They therefore provide a very stiff framework against horizontal thrust from the sliding doors.

Both doors form a semicircle in plan and a quarter-segment of a sphere in three dimensions. In each half, three moving elements slide under one fixed element. Each shell-shaped element is fixed to a hinge at the top of the end arch and is guided horizontally by rails, both tangentially and radially, at the bottom. Each sliding door element has two motor drives, one at each end at ground level.

Opposite: The concrete plinths are the springing points for the arches, and also contain the entrances to the hangar.

The five arches have a structural height of 8 metres (26 feet) and span 225 metres (740 feet), with top chords at 3.4-metre (11-foot) centres and bottom chords at 2-metre (6½-foot) centres. The chords are brace-connected to each other, with the exception of the two bottom chords which are connected by straight members forming a Vierendeel system.

At their ridge the arches are longitudinally connected by a four-chord, 8-metre (26-foot) deep truss, similar in structure to the arches. This ridge beam enables the membrane and the valley cable at the top to be connected, and takes up the large compression force between the two end arches generated by the doors.

Tubular hollow sections were used as structural elements for the arches because of their high torsional resistance and good buckling performance. The chords have an outside diameter of 559 millimetres (22 inches); the diagonals and the bottom straight are 355 millimetres (14 inches) and the side and top straights are 273 millimetres (11 inches).

A PVC-coated membrane was chosen as the roof covering for three reasons: it is light in weight, it has an expected 20-year life and there is experience of its use over 30 years on other types of construction. The stressed membranes span 31 metres (100 feet) clear between the trussed tubular arches in the warp direction and between the ridge truss and the edge cable attached to the arch bases in the fill direction. A prestressed valley cable was placed over the membrane midway between the arches to generate sufficient curvature in the membrane to limit stresses and deflections.

Below: The truss along the spine of the hangar carries a maintenance walkway 100 metres (330 feet) above the floor.

Opposite: Erection of the doors was a major operation.

Initially, all the known and assumed loading data were assembled into one document, which was sent to and checked by the entire project team. After receiving approval, this 'Loading Assumptions Report' formed a very important basis for the whole engineering team carrying out the calculations.

As for special loads, a ±45°C (113°F) temperature variation was applied to the external steelwork and ±10°C (50°F) to the internal steelwork, whilst an additional load from ice up to 30mm (1¼ inches) thick (of 0.21kN/m²/3.5lbf/in²) was applied to the external steel. The influence of foundation settlement (of 50 millimetres/2 inches) was analysed but, as expected, its effect on a structure this size was negligible.

Below: The two-storey office building shows here between the plinths of the hangar's arches.

Opposite: The doors can be seen nesting inside each other when the door is partially open.

The doors

Door weight had a strong influence on costs. The enormous size of each segment – arch length 168 metres (550 feet), bottom width 42 metres (138 feet) – inevitably meant that the doors would be extremely heavy. The upper supports of the doors load the ridge points, which in turn apply concentrated loads to the end arches and therefore influence their section dimensions. The lower support reactions of the doors also dictate the costs of the driving mechanism and the foundations of the doors themselves.

The lightest structure was achieved by adopting the shell principle. The inner part of each door segment comprises a spherical grid of identical horizontal, vertical and diagonal elements, rigidly jointed. The corrugated metal-sheet cladding spans across this grid between the upward-curving side beams, which are 3 metres by 800 millimetres (10 feet by 2½ feet) on plan. Each door segment terminates in a lower beam, 2.3 metres (7½ feet) deep and 800 millimetres (2½ feet) wide. When the doors are opened, the four beams (one on each side of each door) nest concentrically inside each other.

Throughout, the objective of reducing costs by minimizing steel tonnage was maintained. Ultimately, this was achieved by the combined effect of a wide-spanning, lightweight, stressed membrane and thin, light shell structures – plus the knowledge gained from the developments in design, detailing and construction over the previous 60 years. These factors gave the hangar an impressively low weight when compared with its two famous predecessors, at Cardington in the UK and Akron in the USA.

Conclusion

The CargoLifter hangar was designed in 1997/98 and completed in 2000. The smaller CL75 is already being tested and the CL160 was intended to fly in 2003, with commercial operations starting in 2005. However, at the time of writing, CargoLifter AG's operations have halted for financial reasons. Nonetheless, the hangar marks the rebirth of the almost forgotten technological development of airship hangars in Germany, and the airships themselves can be expected to come into their own for heavy, long-distance transport in the 21st century. This hangar, a remarkable feat of engineering, required the execution of bold and imaginative concepts, particularly in the design of the doors. This was developed by Arup in parallel with their design for the opening roof of the Miller Park Stadium in Milwaukee, Wisconsin, USA, which shares a similar structural concept.

Left: The size of the vehicles inside the hangar shows the scale of the structure.

> Millennium Wheel Canal Link

Falkirk, Scotland, UK, 1999–2002

The Forth and Clyde Canal was the first sea-to-sea ship canal in the world, opened in 1773 between Grangemouth on the Forth estuary and Bowling on the Clyde. The Union Canal, built to take barges from the Forth and Clyde at Falkirk to Edinburgh, was completed in 1822. The Forth and Clyde and Union system fell into disuse with the rise in motor-vehicle traffic after the Second World War. In the 1960s the Forth and Clyde was abandoned as a commercial waterway and blocked by new road crossings. Various sections were filled in and bridges and locks fell into disrepair. At Falkirk, where the two canals meet, the Millennium Wheel barge lift enables vessels to pass between the two levels.

In recent years British Waterways, the state organization responsible for canals in the UK, has recognized the value of waterfront development in urban regeneration. Waterfront housing and commercial leisure developments have become extremely popular and can contribute to funding the re-opening of canals for leisure purposes. The cost of reopening the Forth and Clyde canal system was estimated at £78 million and required the removal of over 30 obstructions and the rebuilding and replacement of 120 structures. Grants of £32.2 million were obtained from the lottery-funded Millennium Commission; £18.7 million from Scottish Enterprise, the development agency of the Scottish Assembly; £8.6 million from the European Commission; and £9.3 million from British Waterways. The remaining funds came from local councils and the private sector.

Opposite: The rotor arms of the barge lift create a sculpture as well as an operational engineering structure.

At the junction of the two canals there had originally been a flight of 11 locks, but this was later buried under road construction. To overcome the 35-metre (115-foot) difference in levels, a barge lift was proposed. It was estimated that two barges in each direction would need to be transferred every 15 minutes to cater for the predicted traffic. Tenders were originally sought for a Victorian-style Ferris wheel with four hanging gondolas. However, this was then rejected and a more 'showcase' design sought. Accordingly, following appointment of the contractor, all members of the design and contracting team were brought together for a series of brainstorming sessions to consider alternatives. The brief was for a showpiece wheel befitting the 21st century. Various concepts were considered and rejected, including the use of inclined planes, balanced beams and more-conventional hydraulic lifts.

The project that emerged was for two gondolas, each large enough for two barges, supported by two wheel arms which rotate through 180 degrees for each operation of the lift. These wheel arms have no outer rim and are more like the shape of a cycle spanner than a conventional wheel. The gondolas themselves are supported on rails that run around the inside of holes in each side of the wheel, which rotates

around an axle driven by ten hydraulic motors operating at its aqueduct end. The whole system is counterbalanced and, unlike conventional locks, uses virtually no water because one gondola is raised as the other is lowered. Each gondola carries about 300 tonnes (295 tons) of barge and water.

The arms make for a dramatic architectural and engineering form, reflected in the design of the piers of the upper approach aqueduct. If the gondolas were allowed to travel freely on the rails within the inner rims of the holes in the supporting arms, wheel friction and the inertia of the water mass within a gondola would result in a sideways tilt and there would be a risk of it overturning. It was calculated that after 40 seconds a 4-degree tilt could result in a gondola tipping over. It was therefore necessary to introduce a system to maintain horizontality of the gondolas. Rather than use any damping or motor-controlled system, a mechanical system was developed based on cog wheels of equal diameter fixed to the abutment and each gondola. An intermediate small cog on each side provides the mechanical linkage between the cog fixed to the abutments and those fixed to the gondolas. As the arms rotate, the cog wheels ensure that the gondolas remain in the horizontal plane at all times. This cog-wheel system was initially developed using a model made out of Lego, the construction toy. Because the wheel is finely counterbalanced, the power requirement is low: it is turned by ten 7-kilowatt motors, giving a power cost of just £10 per day.

The engineering concept appears simple, but the scale of the structures presented some difficult engineering problems. Under load the axle bends, causing the ends of the arms of the wheel to move inwards at the top and outwards at the bottom. The arms face repeated 100 per cent stress reversals as the wheel rotates. The gondolas act as simply supported beams, transferring load through rail lines fixed into the inside rim of the circular holes in the arms of the wheel. Various design codes were consulted – including offshore specifications and codes from Norway and Germany – since no codes are directly applicable. The seals and hydraulics had to be designed to cater for the movements resulting from the bending in the structures.

At the lower level of the wheel there is a large basin which allows up to 20 barges to moor while waiting for the lift and also assists in the management of the water regime. The works included a 168-metre (550-foot) long tunnel and a four-span aqueduct at the approach to the upper level of the wheel. Three new locks were required and 2 kilometres (1¼ miles) of new canal.

A tunnel takes the canal under a main rail line and a scheduled ancient monument, the Antonine Wall, which once marked the northern boundary of the Roman Empire. The 100-metre (330-foot) long aqueduct, which forms an integral part of the wheel's architectural impact, has an organic form and was inspired by the spine of a fish skeleton. The architect's wish to have the aqueduct 'float' through the piers, which are at 25-metre (82-foot) centres. This was achieved by supporting the deck on two narrow side sections of column, with

650-tonne (640-ton) vertical loads through each. To confirm that the reinforcement could be fitted into the available space, full-size test panels of the deck-pier interface were cast at ground level.

Shortly prior to the wheel's opening, in a serious act of vandalism, an upstream lock gate was opened and millions of litres of canal water released. This flooded the control circuits, which would have detected and prevented the surge had they been fully commissioned. As a result the Queen was unable to perform the official opening during her tour of Britain to commemorate the 50th year of her reign, but she did visit the project.

Conclusion

Barge lifts were among the first engineering structures built for transport needs that required mechanical as well as civil engineering skills. The use of counterweights, either to pull a barge on rollers up an incline or to lift a tank containing the vessel, is an engineering solution going back to the earliest days of canal building. It involves attaching ropes to both the load to be lifted and the counterweight, and then passing them round a wheel. The rotating barge lift is the modern version of this solution, made possible by the use of a steel rotating arm which can support the bending moment resulting from the cantilevered container. This example is the remarkable result of collaboration between the architect and engineers of many different disciplines. Their imaginative solution to an unusual problem has rapidly become a tourist attraction, and with its eye-catching design it will no doubt be a landmark in Central Scotland for many years to come.

Following spread: The picture shows the interior of Corporation Street Footbridge, Manchester, UK.

Below: The gondolas are kept level by a series of cogs as the wheel turns.

Opposite: The upper level of the canal is carried on a viaduct to the centre of the pool.

Below: This aerial view shows the upper level pool on the right. The lower pool is used as a stopping-off point for tourists travelling along the canal as well as for boats waiting for the lift.

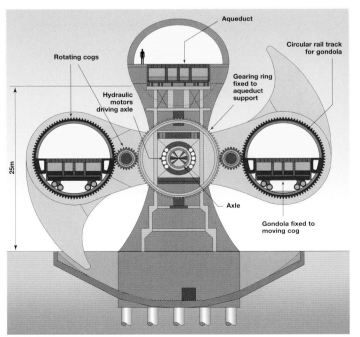

> Appendix

References and Credits

References

Introduction

Ove Arup & Partners 1946–1986, Academy Editions, 1986.
David Dunster (ed.), *Arups on Engineering*, Ernst & Sohn, 1996.
Asa Briggs, *Iron Bridge to Crystal Palace: Impact and images of the industrial revolution*, Thames and Hudson, 1979.
Charles Hadfield, *The Canal Age*, David & Charles, 1968.
Bryan Morgan, *Railways: Civil Engineering*, Longmans, 1971.
Sir John Betjeman, *London's Historic Railway Stations*, 2nd Edition, Capital Transport, 2002.
Nicholas Faith, *The World the Railways Made*, The Bodley Head, 1990.
P. J. G. Ransom, *The Victorian Railway and How it Evolved*, Heinemann, 1990.
Laurence Jankowski de Niewmierzycki (ed.), *Le Patrimoine de la RATP*, Flohic Éditions, 1998.
Michael Hunter and Robert Thorne, *Change at King's Cross*, Historical Publications, 1990.
Caroline Mathieu, *Orsay, from a Station to a Museum*, Éditions Scala, 1999.
G. Drysdale Dempsey, *Tubular and Other Iron Girder Bridges*, Virtue Brothers, London, 1864.
David J. Brown, *Bridges: Three Thousand Years of Defying Nature*, Mitchell Beazley, 1993, rev. 1998.
Philip Jodidio, *Santiago Calatrava*, Taschen, 2001.

Chapter 1: Airports

Terminal 4, John F. Kennedy International Airport, New York, USA
Arup Focus, September 2001, p.1.

Terminal 2, Cologne/Bonn Airport, Germany
E. Trefziger, 'Tough Times', *Passenger Terminal World*, March 2002, pp. 30–40.
M. J. Crosbie, 'The New Modernism of Helmut Jahn', *Architecture Week*, 17 July 2002.

New Terminal, Lester B. Pearson International Airport, Toronto, Canada
N. Doshi and O. Tatcher, 'Fast Forward', *Passenger Terminal World*, June 2002.
ArupNAPA, 'Airport Master Plan Summary Report', Greater Toronto Airport Authority, www.torontoairport.ca, *Arup Focus*, September 2001, p.4.

Hong Kong International Airport, Chek Lap Kok, China
Numerous authors, Hong Kong Airport Core Projects. *The Arup Journal*, 1/1999, pp. 3–60.

Chapter 2: Railways

Hong Kong Airport Railway, China
Numerous authors, Hong Kong Airport Core Projects. *The Arup Journal*, 1/1999, pp. 3–60.

Channel Tunnel Rail Link, UK
M. Bostock and T. Hill, 'Planning High Speed Railways into Europe', *The Arup Journal*, 4/1993, pp. 3–7.
D. Loosemore and N. Shepherd, 'Planning High Speed Railways into Europe: An Update', *The Arup Journal*, 1/1996, pp. 3–5.
N. Lovelace, 'Eastern Promise', CTRL Special Feature, *New Civil Engineer*, 6 June 2002.

St Pancras Station, London, UK
'On to London', CTRL Special Report, *Modern Railways*, October 2000.
A. Knowles, 'St Pancras: No Longer a Terminal Case?', *Rail 432*, 3–16 April 2002, pp. 32–37.
'New and Old', CTRL Supplement, *New Civil Engineer*, 28 June 2001.

King's Cross/St Pancras Underground Station, London, UK
P. Evans, 'King's Cross Underground Station Redevelopment', *Arup Bulletin*, March 2001, p. 16.
H. P. White, *A Regional History of the Railways of Great Britain*, Vol. 3, David & Charles, 1963.
S. Halliday, *Underground to Everywhere – London's Underground Railway in the Life of the Capital*, Sutton Publishing, 2001.
Sheila Taylor, *The Moving Metropolis: A History of London's Transport Since 1800*, Laurence King Publishing, 2001.
'Tube expansion at King's Cross', *Modern Railways*, April 2002, pages 36–37.

Hanging Railway Stations, Wuppertal, Germany
H. Arens and T. Carter, 'The Wuppertal Floating Train', *Elevator World*, December 1997.

Light Rail Stations, Hanover, Germany
R. J. Buckley, 'Hanover: 25 Years of Light Rail Development', *Light Rail Review*.

Chapter 3: Bridges

Øresund Crossing, Denmark/Sweden
J. Nissen, 'The Øresund Link', *The Arup Journal*, 2/1996, pp. 37–41.
K. Falbe-Hansen and J Nissen, 'The Øresund Bridge Completion', *The Arup Journal*, 3/2001, pp. 20–27.

Millennium Bridge, London, UK
T. Fitzpatrick, 'Linking London: The Millennium Bridge', *Royal Academy of Engineering*, June 2001.

Pero's Bridge, Bristol, UK
T. Kerr, 'Pero's Bridge', Arup (unpublished).

Spencer Street Footbridge, Melbourne, Australia
P. Connolly, 'Spencer Street Footbridge, Melbourne, Australia', *The Arup Journal*, 1/2000, pp. 20–22.

Denver Millennium Bridge, Colorado, USA
www.denvermillenniumbridge.com
Architecture Denver
S. Kite and J. Eddy, 'Denver Millennium Bridge', *The Arup Journal*, 1/2003, pp.33–35.

Corporation Street Footbridge, Manchester, UK
S. Clarke, A. Foster and R. Houghton, 'Corporation Street Footbridge, Manchester', *The Arup Journal*, 2/2001, pp. 46–47.

Hulme Arch Bridge, Manchester, UK
N. Hussain, R. Milburn and I. Wilson, 'Hulme Arch Bridge', *The Arup Journal*, 2/1997, pp. 15–17.
N. Hussain and I. Wilson, 'The Hulme Arch Bridge, Manchester', Proceedings of the Institution of Civil Engineers, *Civil Engineering*, 1999, Volume 132, Feb., pp. 2–13. Paper 11640.

Chapter 4: Transport Planning and Special Projects

Transport for the 2000 Olympics, Sydney, Australia
B. Clark, C. Henson and A. Hulse, 'Transportation for the Sydney Olympics'. *The Arup Journal*, 1/2001, pp. 3–8.

International Border Crossings and Transport, Central and Eastern Europe
G. Styles, 'Minimising the Impact of Cross Border Congestion', EU Seminar, Warsaw, September 1999.
E. Humphreys and I. Birch, 'Extending TERFFS to Central and Eastern Europe', *The Arup Journal*, 3/2000, pp. 16–17.

Traffic Calming Scheme, Brooklyn, New York, USA
'Downtown Brooklyn Traffic Calming Project', Final Report, Arup, September 2002.

CargoLifter Airship Hangar, Brand, Brandenburg, Germany
M. Janner, R. Lutz, P. Moerland, and T. Simmonds, 'The CargoLifter Hangar in Brand, Germany', *The Arup Journal*, 2/2001, pp. 24–31.

Millennium Wheel Canal Link, Falkirk, Scotland, UK
D. Hayward, 'Project Study: Falkirk Wheel', *New Civil Engineer*, 18 April 2002.

Picture credits

T=Top B=Bottom L=Left R=Right

All pictures reproduced with kind permission of Arups Photo Library, except for the following. All attempts have been made to contact relevant parties to obtain correct credits.

2 Despang Architekten: 7 Robert Harding Picture Library: 10 Hutchison Library/Bernard Regent: 11 Angelo Hornak: 12 Mary Evans Picture Library: 13 Robert Harding Picture Library: 14 © Paul M.R. Maeyaert: 15 London Transport Museum: 16L Robert Harding Picture Library: 16R Hutchison Library/Robert Francis: 17 Panama Canal Authority: 18 Colin Boocock: 19T AKG London: 19B Robert Harding Picture Library: 20T Private Collection: 20B Tim Pharoah: 21 South American Pictures/Tony Morrison: 22 Hutchison Library/Dirk R.Frans: 23 Robert Harding Picture Library: 24 Ironbridge Gorge Museum Trust/Bridgeman Art Library: 25L QA Photos: 25R Ironbridge Gorge Museum Trust: 26© Paul M.R. Maeyaert: 27 ECE Projektmanagement: 28 Stephen Hodgkiss Design for G-Mex: 29 Hutchison Library/Tony Souter: 30T Hutchison Library/T.Moser: 30B Private Collection: 31 St. Louis Union Station: 32, 33 Robert Harding Picture Library: 34 Philippe Morel: 35 Dennis Gilbert/View: 36 Aviation Photos: 38 Dennis Gilbert/View: 39 Jens Willebrand: 40 Edifice/Schneebels: 41 endex.com/gf/buildings/bbridge/bbridge (Brooklyn Bridge Website and Jim Reiss): 42 Collections/Gena Davis: 43 Honshu-Shikoku Bridge Authority: 44/45 Palladium Photo Design, Koln. Barbara Burg/Oliver Schuh: 47, 49 Frederick Charles: 50L, 51 Skidmore Owings & Merrill: 50RB TAMS Consultants: 52 Frederick Charles: 53, 54 Joe Vericker, Photo Bureau: 55, 56, 57, 58, 59 Frederick Charles: 61, 62, 64 Frank Alexander Rummele, FAR Photographie: 65 Palladium PhotoDesign, Koln: 67, 68, 69, 70, 71 Skidmore Owings & Merrill/Greater Toronto Airports Authority: 74B Foster and Partners: 75, 86 Arup/Gareth Jones: 101 EDAW Earthasia: 104 Chris Gascoigne/View: 105 Terry Farrell & Partners: 117 QA Photos/Rail Link Engineeering: 118 Morgan Est/Studio Edwardo: 119, 121, 122 Union Railways/Rail Link Engineering: 120, 123, 124, 125 QA Photos/Rail Link Engineering: 127 Arcaid/David Churchill: 128, 129, 130 Rail Link Engineering: 131 Camden Local Studies & Archives: 132 Arcaid/Nicholas Kane: 133 Rail Link Engineering: 135, 137, 138, 140, 141 Infraco Sub Service Ltd as Project Managers for London Underground Ltd: 136 London Transport Museum: 143 Torsten Schroeter: 144L AKG London: 149 Jens Willebrand: 150, 151 Despang Architekten: 152, 153 Jens Willebrand: 157 Arup/Jørgen Nissen: 160R Pier Mens: 165 Graham Gaunt: 163, 164, 166L, 167, 168, 169, 170L Grant Smith/View: 173, 174T, 175 Visual Technology, Bristol City Council: 174T Hutchison Library/Andrew Eames: 177, 178, 179, 181 Gollings Photography: 180 Peter Elliott Pty Architects: 184L ,185B Architecture Denver: 183, 185T, 187R Frank Ooms: 189, 190L, 191 Inside Out/Bernard O'Sullivan: 190R Sarah Quill/Bridgeman Art Library: 193, 194, 195 Arup/Peter MacKinven: 196/7 Wilkinson Eyre/Positive Image: 199 Palladium Photo Design, Koln: 201 *Sydney Morning Herald*: 205 Olympic Co-ordination Authority: 211 Colin Boocock, Derby: 213, 214, 215, 216, 217 Charlie Samuels: 219, 220B, 221, 223, 224, 225, 226, 227, 228/9 Palladium PhotoDesign, Koln: 231, 232, 233R Arup/Jim Mackintosh Photography: 233L *New Civil Engineer*: 235 Inside Out/Bernard O'Sullivan.

Project Credits

Chapter 1: Airports

Terminal 4, John F. Kennedy International Airport, New York, USA
Client: Port Authority of New York & New Jersey (Phase 1); JFKIAT LLC (Phase 2)
Design team: TAMS, Skidmore Owings and Merrill LLC, Arup
Arup role: Masterplanning, civil, structural, mechanical, electrical, plumbing, environmental, communications, transportation, acoustic, and fire engineering

Terminal 2, Cologne/Bonn Airport, Germany
Client: Flüghafen Köln/Bonn GmbH
Architect: Murphy/Jahn
Arup role: Structural engineer

New Terminal, Lester B. Pearson International Airport, Toronto, Canada
Client: Greater Toronto Airports Authority
Architects: Airport Architects Canada (Skidmore Owings & Merrill International Ltd, Moshe Safdie Associates Ltd, and Adamson Associates)
Arup role: Masterplanning and conceptual design, structural, mechanical, electrical, fire, and communications engineering

Hong Kong International Airport, Chek Lap Kok, China
The Airport Terminal Building
Client: Airport Authority Hong Kong
Design team: Mott Connell Ltd, Foster and Partners, British Airports Authority plc, Arup
Arup role: Scheme design of superstructure, steelwork structural design, fire and acoustic engineering

Ground Transportation Centre
Clients: Airport Authority Hong Kong/Mass Transit Railway Corporation
Lead consultant: Arup
Architects: Foster and Partners, Anthony Ng Architects Ltd.
Arup role: Civil, structural, bridge, highway, railway, acoustic, fire, maritime, and geotechnical engineering

HACTL SuperTerminal 1
Client and project managers: Hong Kong Air Cargo Terminals Ltd
Lead consultant: Arup
Architectural consultant: Foster and Partners
Arup role: Civil, structural, geotechnical, mechanical, and electrical engineering, building project management, supervision, and testing/commissioning

Lufthansa Catering Facility
Client: Lufthansa Sky Chefs
Architect: Llewelyn-Davies Asia
Arup role: Project manager and design checking engineer

Chapter 2: Railways

Hong Kong Airport Railway, China
Hong Kong Station
Client: Mass Transit Railway Corporation
Architect: Arup Associates in association with Rocco Design Partners
Arup role: Civil, structural, geotechnical and acoustic engineering

Kowloon Station
Client: Mass Transit Railway Corporation
Architect, masterplanner and lead consultant: Terry Farrell & Partners
Collaborating architects: Ho & Partners
Arup role: Civil, structural, geotechnical, acoustic, traffic, and transportation engineering

Tsing Yi Station
Client: Mass Transit Railway Corporation
Station
Architect and lead consultant: Wong Tung and Partners
Arup role: Civil, structural, geotechnical and acoustic engineering
Viaducts
Lead consultant: Arup
Arup role: Civil and geotechnical engineering

Tung Chung Station
Client: Mass Transit Railway Corporation
Lead consultant: Arup
Architects: MTRC Architects Department in association with Rocco Design Partners
Arup role: Civil, structural, geotechnical and acoustic engineering

Channel Tunnel Rail Link, UK
Client: London & Continental Railways (Arup, Bechtel Ltd, SG Warburg & Co Ltd, National Express Group PLC, Systra, London Electricity PLC, Halcrow) Designers: Rail Link Engineering (Arup, Bechtel Ltd, Halcrow, and Systra)
Arup role: Original route concept, planning, environmental, tunnel, geotechnical, route civil, construction, rail safety engineering, and community relations

St Pancras Station, London, UK
Client: London & Continental Railways
Designers: Rail Link Engineering
Arup role: Planning, civil, structural, mechanical, electrical, public health, infrastructure, geotechnical, environmental, transportation, and acoustic engineering

King's Cross/St Pancras Underground Station, London, UK
Client: London Underground Ltd
Project manager: Infraco Sub-Surface Ltd
Lead designer: Arup
Architect: Allies and Morrison
Arup role: Design management, construction planning/programming, quantity surveying, health & safety, civil, structural, building services, electrical, public health, communications, fire, geotechnical, tunnels, traffic, environmental, acoustic, and value engineering

Hanging Railway Stations, Wuppertal, Germany
Client: Wuppertaler Stadtwerke AG
Architects: Jaspert and Steffens, Cologne, Schuster Architekten, Düsseldorf, Claudia Drosdowski, Radevormwald, and Chamier & Molina, Düsseldorf.
Arup role: Structural, services, and communications engineering

Light Rail Stations, Hanover, Germany
Client: TransTec Bauplanungs- und Managmentgesellschaft GmbH
Architect: Despang Architekten, Hanover
Arup role: Structural engineering

Chapter 3: Bridges

Øresund Crossing, Denmark/Sweden
Client: Øresundsbro Konsortiet
Designer: ASO Group (Arup, SETEC Travaux Publics et Industriels, Gimsing and Madsen A/S, ISC Consulting Engineers A/S) with Georg Rotne Architect MAA
Contractor: Sundlink (Skanska, Hochtief, Højgaard & Schulz, Monberg & Thorsen)
Contractor's engineer: CV Joint Venture (COWIconsult, VBB Viak)
Arup role: Design concept, geotechnical, civil, structural and fire engineering, risk assessment, preparation of tender documents, construction monitoring, and quality assurance audits.

Millennium Bridge, London, UK
Client: London Borough of Southwark
Designers: Arup, Foster and Partners, Sir Anthony Caro
Arup role: Planning supervision, project management, geotechnical, structural, civil, aerodynamic, maritime, lighting, damping and acoustic engineering, and archaeology

Pero's Bridge, Bristol, UK
Client: Bristol City Council
Lead designer: Arup
Artist: Ms Eilis O'Connell
Arup role: Project planning, geotechnical, civil and structural engineering

Spencer Street Footbridge, Melbourne, Australia
Client: Melbourne Convention and Exhibition Centre Trust
Project Manager: Office of Major Projects
Architect: Peter Elliott, P/L Architects
Arup role: Structural, civil and façade engineering

Denver Millennium Bridge, Colorado, USA
Client: Central Platte Valley Metropolitan District
Architect: ArchitectureDenver
Arup role: Civil and structural engineering, operation and maintenance manual, site supervision

Corporation Street Footbridge, Manchester, UK
Client: Manchester Millennium Ltd
Owner: Prudential Portfolio Managers Ltd
Lead designer: Arup
Architect: Hodder Associates
Arup role: Civil, structural, electrical and façade engineering

Hulme Arch Bridge, Manchester, UK
Client: Hulme Regeneration Ltd
Architects: WilkinsonEyre
Arup role: Civil, structural, geotechnical, highway and lighting engineering, and contract administration

Chapter 4: Transport Planning and Special Projects

Transport for the 2000 Olympics, Sydney, Australia
Clients: Olympic Co-Ordination Authority and Olympic Roads and Transport Authority
Arup role: Rail, road and pedestrian transport strategy and modelling, crowd safety, communications and fire engineering

International Border Crossings and Transport, Central and Eastern Europe
Client: European Commission/'PHARE' Programme
Border crossings study team leader: Arup
Other border crossings study team member: HM Customs International Assistance
Arup role: International transportation advice

TERFFS study team leader: Arup
Other TERFFS team members: Prognos, Regional Consulting, Transman, DHV (CR), DHV (BV), Railplan
Arup role: Project management, economic analysis, forecasting, rail freight institutional matters

Traffic Calming Scheme, Brooklyn, New York, USA
Client: New York City Department of Transportation
Lead consultant: Arup
Urban designer: Jambekhar Strauss
Landscape architect: Judith Heinz Associates
Arup role: Traffic calming study, data collection, evaluation of alternatives; followed by design, implementation, monitoring and testing of pilot programme

CargoLifter Airship Hangar, Brand, Brandenburg, Germany
Client: CargoLifter AG, Wiesbaden
Architect and masterplanner: SIAT Architektur + Technik, München
Arup role: Structural engineering

Millennium Wheel Canal Link, Falkirk, Scotland, UK
Client: British Waterways Scotland
Design and build contractor: Morrison Construction/Bachy Soletanche JV
Architect: RMJM Ltd.
Wheel designer/fabricator: Butterley Engineering Ltd
Arup role: Lead civil engineer for canals, aqueducts, locks, wheel substructure and support building, and all other site infrastructure; structural, mechanical and electrical engineer for visitor centre.

Index

Numbers in *italics* refer to illustrations.

Adamson Assoc./Moshe Safdie & Assoc./
SOM: Lester B. Pearson International
Airport, Toronto, new terminal 66, *67–71*, 68–71
Adler Brücke station, Wuppertal, Germany
(Schuster Architekten) *146*
Aeroports de Paris: Paris Charles de Gaulle
airport, terminals 39
aesthetics, and bridges 16
Airport Express Line (AEL), Hong Kong 72, 90
airports 36–9
    see also individual airports
airships 218
    see also CL160 airship
Almeda Bridge, Valencia, Spain (Calatrava) *44–5*
Alsop, Will 35
Amsterdam:
    Centraal station 26
    Schiphol Airport 36
'archi-neering' 60, 64
Architectural Design 10
ArchitectureDenver: Denver Millennium Bridge,
Colorado 182, *183–7*, 184, 185
Arup, Sir Ove 8, 9, 10
Arup, Ove/Arup GmbH/ArupNAPA/Arup
Transportation/Ove Arup & Partners:
    bus station, Dublin 10
    Brooklyn Traffic Calming Project, New York 214
    CargoLifter Hangar, Brandenburg,
    Germany 218, 229
    Central European border programme 206–11
    Channel Tunnel Rail Link, south-east England
    116, 118–19
    Chek Lap Kok Airport, Hong Kong 74, 87
    Cologne/Bonn Airport, Terminal 2 60
    Hanging Railway (Wuppertaler Schwebebahn), Wuppertal,
    Germany 144
    Hong Kong Station 93
    Hulme Arch Bridge, Manchester 192
    Kingsgate Footbridge, Durham 10
    Kowloon Station 103
    Lester B. Pearson International Airport,
    Toronto 66, 71
    Millennium Bridge, London 165, 168, *169*, 171
    Miller Park Stadium, Milwaukee, retractable
    roof 220, 229
    Øresund Crossing, Denmark/Sweden 158
    Pero's Bridge, Bristol 172
    sketched options for shell structures *9*
    Spencer Street Footbridge, Melbourne 179
    taxi rank designs 99
Ashford International station, Kent 121
ASO group: Øresund Crossing, Denmark/Sweden
16, 156, *157–61*, 158, *160–1*
Auckland Harbour Bridge, New Zealand 171

Bangkok: Suvarnabhumi Airport 39
Barlow, W.H.: train shed, St Pancras Station, London
128, *130–1*, 131, 133, *133*
bascule bridges *174*
Beck, Harry: London Underground map *15*
Betjeman, John 10, 12
Bilbao:
    Bilbao Airport (Calatrava) 39
    Guggenheim Museum (Gehry) 23
Birmingham: Link Bridge, National Exhibition
Centre 170

Bombay, India: Victoria Terminus station
(*later* Chhatrapati Sivaji) *32*, 33
Bondi Beach, Sydney *204*
Border Crossings, International, Central Europe,
EU programme (Arup) 206, *207–9*, 208–11
Brand, Brandenburg: CargoLifter Hangar (SIAT
Architektur+Technik) 218, *219–21*, 220, 222,
223–9, 225–7, 229
Brassey, Thomas 125
Bridge of Sighs, Venice *190*, 191
bridges 40–3
    and aesthetics 16
    see also individual bridges
Bristol:
    Clifton Suspension Bridge (I.K.Brunel) 41
    Floating Harbour 172
    Pero's Bridge (O'Connell) 172, *173–5*, 174–5
British Rail (BR) 116, 119
Brooklyn Bridge, New York (Roebling) 41, *41*
Brooklyn, New York:
    Brooklyn/Queens Expressway 212
    Brooklyn Traffic Calming Project (Arup)
    212, *213–17*, 214–17
    Gowanus Expressway 212
Brunel, Isambard Kingdom 25, 125
    Clifton Suspension Bridge, Bristol 41
    Royal Albert Bridge, Saltash, near
    Plymouth *24*, 41
Brunel, Marc: Thames Tunnel, Rotherhithe,
London 25, *25*
Burkhardt, Berthold 220, 222

Calatrava, Santiago 42
    Almeda Bridge, Valencia, Spain *44–5*
    Bilbao Airport, Spain 39
    Grand Canal, Venice, bridge 43
    Lyon Satolas airport, station 34
Calcutta: Howrah Bridge *7*
canals 17, 24
Cardington, near Bedford, airship hangar 227
CargoLifter AG 218
CargoLifter Hangar, Brandenburg, Germany (SIAT
Architektur+Technik) 218, *219–21*, 220, 222, *223–9*,
225–7, 229
Caro, Sir Anthony (with Foster): Millennium Bridge, London
16, *156*, 162, *163–71*, 165–6, 168, *170–1*
Central and Eastern European Countries (CEECs):
EU rail-freight concept 210–11
Central Europe:
    EU border infrastructure study (Arup) 208–9
    EU border-crossings programme (Arup) 209–10
Central Station, Hong Kong 90
Chamier & Molina Architekten: Hammerstein station,
Wuppertal, Germany *147*
Channel Tunnel 25, *25*
Channel Tunnel Rail Link (CTRL) (RLE), south-east England
33, 116, 118–25, *118–25*
Charles de Gaulle airport, Paris, terminals (Aeroports
de Paris) 39
Charlton, Sir Bobby 196
Charter, Anthony 82
Chek Lap Kok Airport, Hong Kong 72
    Ground Transportation Centre (Foster and Partners,
    Anthony Ng Architects Ltd) *73*, *78–81*, 80–1, 92
    HACTL SuperTerminal (Foster and Partners) 1 82, *82–3*,
    84, *85*
    Lufthansa Catering Building (Llewelyn-Davies Asia)
    86–7, *86–7*
    Terminal Building (Foster and Partners)
    74, *74–7*, 76–7

Chester: Groves Suspension Bridge 171
Chris Wilkinson Architects: Hulme Arch Bridge,
Manchester 192, *193–7*, 194, 196
CIS (Commonwealth of Independent States) 210
city regeneration 23–4
CL160 airship 218, 220, 229
Clayden, James 180
Clifton Suspension Bridge, Bristol (I.K.Brunel) 41
Coalbrookdale, Shropshire: Ironbridge 40, *40*
Cologne/Bonn Airport, Germany, Terminal 2 (Murphy/Jahn)
60, *61–5*, 62–4
Colt, Henry 40
Commonwealth of Independent States (CIS) 210
computers 25, 57
concrete shell structures (Ove Arup) 10
Corporation Street Footbridge, Manchester (Hodder
Associates) 42, 188, *189–91*, 190–1, *235*
counterweight principle *174*, 233
Crosbie, Michael J., *Murphy/Jahn: Six Works* 60
CTRL *see* Channel Tunnel Rail Link
Curitiba, Brazil: triple-articulated buses *21*

Darby, Abraham 40
Dartford Bridge, Kent 162
Denton Corker Marshall Architects: Melbourne Exhibition
Centre, Australia 176
Denver:
    Denver International Airport 39
    Millennium Bridge (ArchitectureDenver) 182,
    183–7, 184, 185
    pedestrian mall, 16th Street (Pei) 182
design flexibility 37
design team concept 8
Despang Architekten: light rail stations, Hanover, Germany
148, *149–53*, 152–3
Dubai Airport 36
Dublin: bus station (Scott) 10
Durham: Kingsgate Footbridge (Arup) 10

Ebbsfleet International station, London 118, 120, 122, *122*
Eiffel, Gustave 25
Eindhoven, Netherlands: traffic-calming project *20*
English Heritage 131, 138
environmental works 125
European Conference of Ministers of Transport 210
European Passenger Services (EPSL) 118
European Union (EU):
    Central European border-crossings programme
    (Arup) 206–11
    PHARE programme 208
    Trans European Networks 158
    Trans European Rail Freight Freeway
    (TERFF) 210–11
Eurorail consortium 116, 119
Eurostar trains 116, *117*, 128, 130
Euston Station, London 12, *12*, 26

Falkirk, Scotland: Millennium Wheel Canal Link (RMJM)
230, *231–3*, 232–3
Farley building, New York (McKim, Mead & White) 28, *30*, 33
Fédération Internationale des Ingénieurs-Conseils (FIDIC)
25, 209-10
Ferguson, Sir Alex 196
Financial Times 162, 165
flexibility of design 37
flight kitchens 86
Floating Harbour, Bristol 172